Current Topics in Microbiology and Immunology

189

Editors

Cytotoxic T-Lymphocytes in Human Viral and Malaria Infections

Edited by M. B. A. Oldstone

With 13 Figures and 21 Tables

Springer-Verlag
Berlin Heidelberg New York
London Paris Tokyo
Hong Kong Barcelona
Budapest

Michael B.A. Oldstone

Department of Neuropharmacology
Division of Virology

The Scripps Research Institute
10666 North Torrey Pines Road
La Jolla, California 92037
USA

Cover illustration: See page 4, Fig. 1a.

Cover design: Harald Lopka, Ilvesheim

ISSN 0070-217X
ISBN 3-540-57259-7 Springer-Verlag Berlin Heidelberg New York
ISBN 0-387-57259-7 Springer-Verlag New York Berlin Heidelberg

Library of Congress Catalog Card Number 15-12910
Printed in Germany

Typesetting: Thomson Press (India) Ltd, New Delhi
SPIN 10096435 27/3020-5 4 3 2 1 0 – Printed on acid-free paper.

Contents

List of Contributors

(Their addresses can be found at the beginning of their respective chapters.)

The Role of Cytotoxic T Lymphocytes in Infectious Disease: History, Criteria, and State of the Art

M. B. A. OLDSTONE

Observations made on children with genetic deficiencies of their immune system reveal that those lacking the capacity to make antibodies nevertheless handle most viral infections as well as normal individuals do. Such a- or hypogammaglobulinemic children are often susceptible to bacterial infections. Conversely, children with genetic deficiencies in their ability to mount cell-mediated immune responses or those who acquire diseases of lymphoid tissues later in life are often susceptible to a range of viral infections (reviewed in GOOD 1991; RICHES 1992).

The immune system consists of humoral (antibody) and cellular (lymphocyte, monocyte, macrophage) responses that most often function synergistically to offer the host protection from invading microbes. Immunization, primarily developed and used most successfully against agents causing acute infection in humans and domestic animals, has been focused toward raising and enhancing antibody responses to glycoprotein or structural protein antigens present on surface virions, although cytotoxic T lymphocytes (CTL) also play major roles in the control of acute infection. Protection from persistent infection is more complex. The viruses are cell associated and often viral surface glycoprotein expression is downregulated (BUCHMEIER and WELSH 1979; OLDSTONE and BUCHMEIER 1982; LIPKIN et al. 1989). In this scenario, to act against virus infected cells, antibodies are not efficient for lysing infected cells. Several million antibody molecules are needed, along with effector molecules of the complement system, to destroy virally infected cells—an ineffective system (SISSONS et al. 1979, 1980).

To better handle persistent infection, the organism's preference is towards the other effector arm of the immune response, consisting of lymphocytes, which detect very low levels of viral antigen on surfaces of infected cells. These lymphocytes are cytotoxic for virally infected cells. Indeed, T cells and CTL by inference are believed to require no more than 100 viral protein (peptide) molecules (DEMOTZ et al. 1990) for activation. CTL can recognize viral sequences expressed on cells that antiviral antibodies are unable to detect. CTL are effective against nonglycosylated immediate–early or early proteins of a virus that are transcribed many hours before structural viral

Viral Immunobiology Laboratory, Division of Virology, Department of Neuropharmacology, The Scripps Research Institute, La Jolla, CA 92037, USA

Current Topics in Microbiology and Immunology, Vol. 189

proteins are made and, most importantly, prior to assembly of the virus into an infectious unit (reviewed in OLDSTONE 1991). Further, CTL recognize infected cells as foreign and destroy them during the latent period of virus infection when the late-acting genes (that encode structural viral proteins) are blocked. CTL can effectively and efficiently recognize and lyse a virally infected cell expressing regulatory proteins (i.e., Nef of human immunodeficiency virus, HIV, immediate–early gene product of cytomegalovirus, CMV) and proteins of the replicative complex (i.e., nonstructural, nuclear protein of influenza virus, Pol and Gag of HIV), all usually made early in the infectious cycle. The strategic advantage to the host in its battle with viruses is that the CTL effector arm can eliminate potential factories (cells) before they produce a finished infectious product (ZINKERNAGEL and ALTHAGE 1977; OLDSTONE 1991).

Now, in 1993, it is abundantly clear that virus-specific CTL are generated in most, if not all, infections by human RNA and DNA viruses. This volume is dedicated towards collecting and disseminating this information. However, the role played in controlling human infection is often circumstantial unless it occurs in genetically deficient or lymphoid-depleted individuals. Yet a plethora of experimental animal studies in which CTL are chemically or genetically deleted combined with reconstitution studies with CTL clones attest to their commanding role in the control of several viral infections, including those with lymphocytic choriomeningitis virus (LCMV), influenza virus, respiratory syncytial virus, CMV infections as well as an important role in rabies and herpes simplex virus infections. Even in the LCMV model, which is acknowledged to be controlled by CTL activity, antibodies can be shown to play a role. For example, CTL reduce viral titers by 4–5 logs (BYRNE and OLDSTONE 1984) and antibodies by 1–2 logs (BALDRIDGE and BUCHMEIER 1992). Interestingly, the murine model of CMV and human CMV point not only to effectiveness of CTL, but a role for NK cells and antibody (BIRON et al. 1989; KOSZINOWSKI et al. 1990). While focusing on CTL, it is nevertheless important to remember the synergistic role of other participants in the immune response.

How do CTL work? CTL recognize proteolytic fragments of viral proteins that are presented at the cell surface by major histocompatibility complex (MHC) glycoprotein molecules (ZINKERNAGEL and DOHERTY 1974; TOWNSEND et al. 1986). MHC molecules are divided into class I and class II, with class I molecules recognized by a CTL subset that bears the CD8 surface marker and class II molecules recognized by CTL or T-helper cells that bear the CD4 surface marker. MHC class I utilizes primarily a cytosolic pathway. MHC class I is found on nearly all cells in the body, an exception being neurons (JOLY et al. 1991; JOLY and OLDSTONE 1992). CTL recognize viral peptides bound to the MHC glycoprotein. The bound peptide sequence is linear and occurs as a consequence of proteolytic fragmentation of a viral protein usually synthesized within the cell. The number of peptides per viral protein able to complex with a MHC glycoprotein in a manner that allows CTL recognition is limited and ranges from one to generally no more than three peptides per viral protein, although recent observations with HIV suggest a larger number of epitopes per viral proteins for this infection (WALKER and PLATA 1990). The size of the peptide has been mapped experimentally and consists optimally of nine to 12 amino acids (reviewed in OLDSTONE 1991), although lower affinity recognition with as few as five amino acids has been noted (WHITTON et al. 1989). Mapping of epitopes utilized recombinant technology, overlapping peptides, and single amino acid truncations from the amino or carboxy termini to initially decode the minimal and optimal peptide size required for CTL recognition as well as recording functional lysis of target cells. Recently the MHC–peptide complex has been directly isolated and chemically identified from virally infected cells (ROTZSCHKE et al. 1990; VAN BLEEK and NATHANSON 1990).

There are a number of milestones in the tale of deciphering CTL activity and structure. The first occurred in 1968, when Brunner and Cerottini, working in Lausanne, Switzerland, reported the use of the quantitative chromium-51 release assay as a marker of membrane injury after cell–cell interaction (BRUNNER et al. 1968). The utility of the assay to study lymphocyte–target cell interactions was appreciated by Cerottini and his colleagues (reviewed in CEROTTINI 1993). Shortly thereafter, Cerottini came as a postdoctoral fellow to Scripps in La Jolla, and on his urging I tested and observed that lymphocytes from mice primed with LCMV lysed (^{51}Cr release) syngeneic LCMV-infected murine targets, but not targets infected either with an indifferent virus or target cells of monkey origin infected with LCMV (OLDSTONE et al. 1969; OLDSTONE and DIXON 1971). Concurrently, Lundstedt in Morgens Volkert's laboratory in Copenhagen, also working with LCMV, found roughly equivalent results (LUNDSTEDT 1969). At this time Martin Raff in Avrion Mitchinson's laboratory in London began utilizing antibody to Th1.2 to segregate T (thymus)-derived from B (bone marrow)-derived lymphocytes, and this reagent was then utilized by Cole, Nathanson, and Pendergast in Baltimore to document that the killer lymphocytes in LCMV mediating cell injury bore the Th1.2 marker (COLE et al. 1972). Next, the seminal observation by ZINKERNAGEL and DOHERTY (1974) in Canberra, Australia, opened wide this field by documenting the requirement for MHC restriction in CTL recognition of the target cell. This contribution also pointed to a role for MHC transplantation antigens and suggested a selective advantage to the host for MHC diversity. Thereafter, many laboratories documented and continue to show CTL killing with MHC restriction and virus specificity for many viruses. In the late 1970s and early 1980s, Askonas (LIN and ASKONAS 1981) in London and Braciale (LUKACHER et al. 1984) in St. Louis, by obtaining and using CTL clones, allowed a more rigorous analysis of CTL recognition and activity. Alain Townsend, Andrew McMichael, and their colleagues provided the next advance when they reported that a peptide from the protein and not the whole viral protein was bound to the MHC for CTL recognition (TOWNSEND et al. 1986). Thereafter, Bevan in La Jolla showed that microinjected cytoplasmic peptide was presented to MHC (MOORE

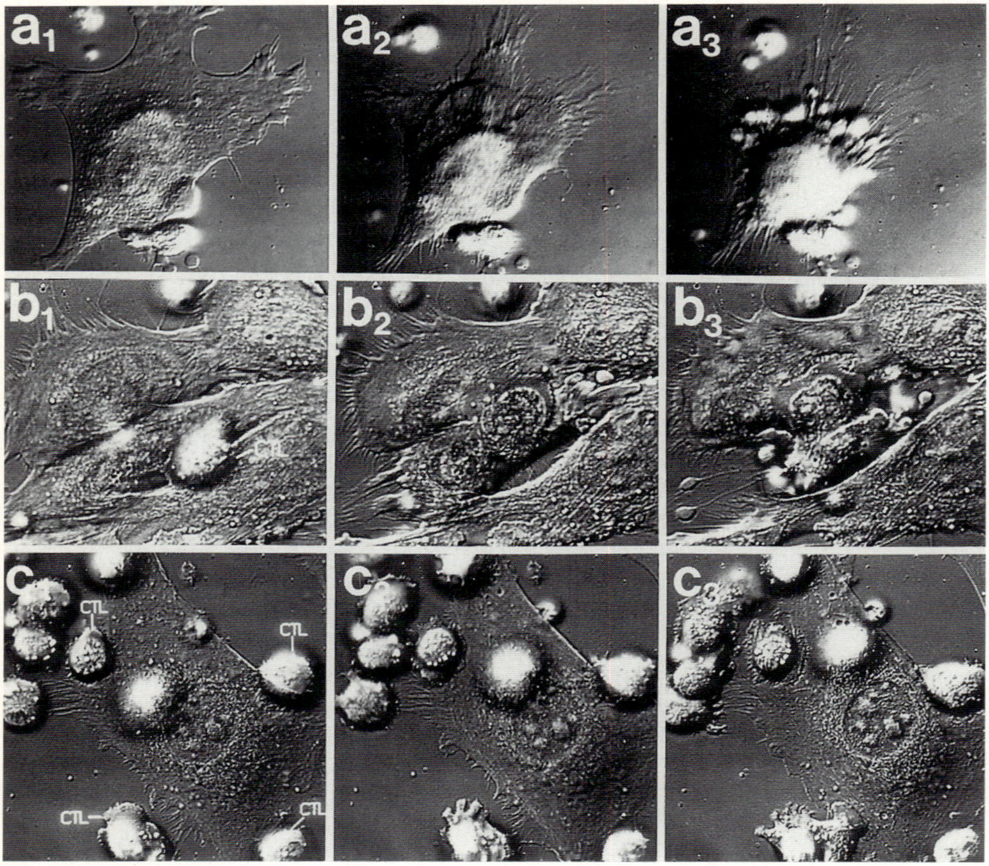

Fig. 1a–c. Photomicrographs documenting disordered morphology of virally infected target cells being attacked and killed by major histocompatibility complex (MHC)-restricted viral-specific cytotoxic L lymphocytes (*CTL*). Panel $\mathbf{a_1}$ shows a lymphocytic choriomeningitis (LCMV)-specific CTL binding to a MHC-restricted target cell. Within 10–20 s of binding, cytoplasmic granules of the CTL reorient towards and flow to the CTL–target cell interface. Panels $\mathbf{a_2}$ and $\mathbf{a_3}$ show the kinetics of killing that take place within 2–3 min. Panels $\mathbf{b_{1-3}}$ show the same effect, but after the viral peptide is microinjected into the cytoplasm of the target cell (the cell in the center of the three targets displayed). Note the CTL lying on top of the target cell. Panels $\mathbf{c_{1-3}}$ by contrast display the interactions of syngeneic CTL and the target cell, but the target cell is expressing an incorrect viral peptide. CTL attach but their cytoplasmic granules fail to reorient towards the target, and the target cell is not killed even after 30 min of observation. Similar findings are seen when the correct peptide is presented by the target cell but to an allogeneic virus-specific CTL. (Unpublished data from K. Hahn, R. DeBiasio, D. Lansing Taylor, and M.B.A. Oldstone, manuscript in preparation, 1993)

et al. 1988), while Whitton, also in La Jolla (WHITTON and OLDSTONE 1989), showed similar results for endogenous processing of a viral minigene encoding the CTL peptide. The conclusion of the first act in this grand theater was the solving of the three-dimensional structure of the MHC class I molecule by Bjorkman, Wiley, Strominger, and colleagues at Harvard (BJORKMAN et al. 1987), revealing two alpha helices bordered by a floor of beta sheaths, the peptide residing in the helices. In concert, Mak and colleagues in Toronto and Davis at Stanford (YANAGI et al. 1985; DAVIS et al. 1986) defined the molecular structure of the T cell receptor that recognizes the peptide–MHC complex. From this brief review, the contributions made by an international group of scientists are evident. Recently, direct isolation of viral peptide, processed by MHC glycoprotein from the infected cell by Nathanson's (VAN BLEEK and NATHANSON 1990) and Rammensee's (ROTZSCHKE et al. 1990) laboratories has been accomplished. Use of this and other data by FALK et al. (1991) provided insights into motifs of peptides bound to MHC molecules. Crystallographic study of different peptides restricted by the same MHC molecule by Wilson's and Peterson's laboratories (FREMONT et al. 1992; MATSUMURA et al. 1992) have enhanced our understanding of structure–function relationships of MHC–peptide interactions.

Once the MHC glycoprotein–peptide complex is on the cell surface, it is recognized by a CTL according to the latter's receptor, composed of alpha and beta chains. If not recognized, the MHC complex likely dissociates, MHC glycoproteins recycle, and the peptide is degraded. The molecular and cellular biology of the molecules involved in CTL attraction to the target, flow and penetration of CTL granules, mechanism of target cell disorder, and death are all current topics of investigation, and several of these points are revealed in photographs from the study of Klaus Hahn and colleagues (Fig. 1).

The MHC heavy chain and its associated light chain (β_2-microglobulin) are encoded by different genes. Their synthesis and assembly occur at the endoplasmic reticulum (ER), and it is currently believed that the assembled MHC molecule does not migrate to the cell surface until a specific peptide binds within the MHC α_1 and α_2 domains (Fig. 2). During this scenario, the β_2-microglobulin is thought to assist in locking the peptide within the MHC groove, causing a change in conformation that allows the complexed molecule to leave the ER. With the exception of neurons (JOLY et al. 1991), other cells studied so far transcribe β_2-microglobulin and the heavy (h) chain of MHC coordinately. Neurons transcribe β_2 normally, but are believed to be deficient in transcription of the MHC h-chain.

This volume presents data from those who have made significant contributions to defining CTL activity to a variety of human viruses and to the protozoa that cause malaria. Further, this volume establishes a foundation for considering CTL activity in several human infections. Research interest in several laboratories for the future is being focused on (a) the correlation of CTL activity in humans with protection and disease and (b) the generation

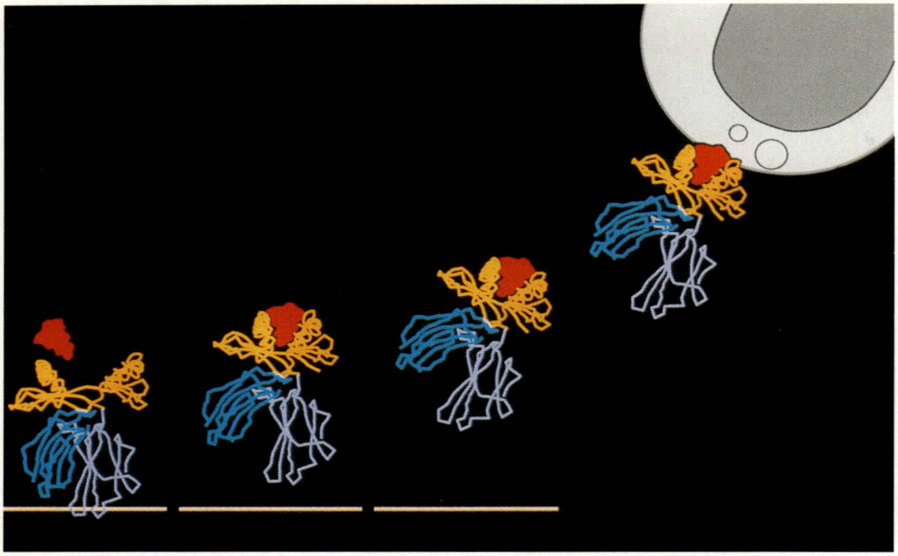

Fig. 2. A cartoon showing sequentially a peptide derived from a viral protein binding within the α_1–α_2 helices of the major histocompatibility complex (MHC) glycoprotein, the peptide being locked into place by the β_2-microglobulin and the MHC–peptide complex traveling to the cell surface, where it is to be surveyed by a cytotoxic T lymphocyte (CTL)

and possible use of adoptive therapy of CTL to modulate and treat human infections and cancers.

Acknowledgments. This is Publication Number 8056-NP from the Division of Virology, Department of Neuropharmacology, The Scripps Research Institute, La Jolla, CA. This work was supported in part by USPHS grants AI-09484, NS-12428, AG-04342, and MH-47680.

References

Baldridge JR, Buchmeier MJ (1992) Mechanisms of antibody-mediated protection against lymphocytic choriomeningitis virus infection: mother-to-baby transfer of humoral protection. J Virol 66: 4252–4257

Biron CA, Byron KS, Sullivan JL (1989) Severe herpesvirus infections in an adolescent without natural killer cells. N Engl J Med 320: 1731–1735

Bjorkman PJ, Saper MA, Samraoui B, Bennett WS, Strominger JL, Wiley DC (1987) Structure of the human class I histocompatibility antigen, HLA-A2. Nature 329: 506–511

Brunner KT, Mauel J, Cerottini JC, Chapuis B (1968) Quantitative assay of the lytic action of immune lymphoid cells on ^{51}Cr-labelled allogeneic target cells in vitro: inhibition by isoantibody and by drugs. Immunology 14: 181–196

Buchmeier MJ, Welsh RM (1979) Protein analysis of defecting interfering lymphocytic choriomeningitis virus and persistently infected cells. Virology 96: 503–515

Byrne JA, Oldstone MBA (1984) Biology of cloned cytotoxic T lymphocytes specific for lymphocytic choriomeningitis virus: clearance of virus in vivo. J Virol 51: 682–686

Cerottini J-C (1993) T-cell-mediated cytotoxicity: a historical note. In: Sitkovsky MV, Henkart PA (eds) Cytotoxic cells: recognition, effector function, generation, and methods (in press)
Cole GA, Nathanson N, Prendergast RA (1972) Requirement for theta-bearing cells in lymphocytic choriomeningitis virus-induced central nervous system disease. Nature 238: 335–337
Davis MM, Lindsten T, Gascoigne NR, Goodnew C, Chien YH (1986) T cell receptor gene structure and function. Cell Immunol 99: 24–28
Demotz S, Grey HM, Sette A (1990) The minimal number of class II MHC-antigen complexes needed for T cell activation. Science 249: 1028–1030
Falk K, Rotzschke O, Stevanovic S, Jung G, Rammensee HG (1991) Allele-specific motifs revealed by sequencing of self-peptides eluted from MHC molecules. Nature 351: 290–296
Fremont DH, Matsumura M, Stura EA, Peterson PA, Wilson IA (1992) Crystal structures of two viral peptides in complex with murine MHC class I H-$2K^b$. Science 257: 919–927
Good RA (1991) Experiments of nature in the development of modern immunology. Immunol Today 12: 283–286
Joly E, Oldstone MBA (1992) Neuronal cells are deficient in loading peptides onto MHC class I molecules. Neuron 8: 1185–1190
Joly E, Mucke L, Oldstone MBA (1991) Viral persistence in neurons explained by lack of major histocompatibility complex class I expression. Science 253: 1283–1285
Koszinowski UH, Val MD, Reddehase MJ (1990) Cellular and molecular basis of protective immune response to cytomegalovirus infection In: McDougall JK (ed) Cytomegaloviruses. Springer, Berlin Heidelberg New York, pp 189–220 (Current topics in microbiology and immunology, vol 154)
Lin Y, Askonas BA (1981) Biological properties of an influenza A virus specific killer T cell clones. Inhibition of virus replication in vivo and induction of delayed hypersensitivity reactions. J Exp Med 154: 225–234
Lipkin WI, Villarreal LP, Oldstone MBA (1989) Whole animal section in situ hybridization and protein blotting: new tools in molecular analysis of animal models for human disease. In: Haase AT, Oldstone MB (eds) In situ hybridization. Springer, Berlin Heidelberg New York, pp 33–54 (Current topics in microbiology and immunology, vol 143)
Lukacher AE, Braciale VL, Braciale TJ (1984) In vivo effector function of influenza virus-specific cytotoxic T lymphocyte clones is highly specific. J Exp Med 160: 814–826
Lundstedt C (1969) Interaction between antigenically different cells: virus-induced cytotoxicity by immune lymphoid cells in vitro. Acta Pathol Microbiol Scand 75: 139–152
Matsumura M, Fremont DH, Peterson PA, Wilson IA (1992) Emerging principles for the recognition of peptide antigens by MHC class I molecules. Science 257: 927–934
Moore MV, Carbone FR, Bevin MJ (1988) Introduction of soluble protein into the class I pathway of antigen processing and presentation. Cell 54: 777–785
Oldstone MBA (1991) Molecular anatomy of viral persistence. J Virol 65: 6381–6386
Oldstone MBA, Buchmeier MJ (1982) Restricted expression of viral glycoprotein in cells of persistently infected mice. Nature 300: 360–362
Oldstone MBA, Dixon FJ (1971) The immune response in lymphocytic choriomeningitis viral infection. In: Miescher P (ed) 6th international symposium of immunopathology. Schwabe, Basel, pp 391–398
Oldstone MBA, Habel K, Dixon FJ (1969) The pathogenesis of cellular injury associated with persistent LCM viral infection. Fed Proc 28: 429
Riches PG (1992) Viral infections complicating immunodeficiencies. Clin Ther 140: 123–129
Rotzschke O, Falk K, Deres K, Schild H, Norda M, Melzger J, Jung G, Rammensee H-G (1990) Isolation and analysis of naturally processed viral peptides as recognized by cytotoxic T cells. Nature 348: 252–254
Sissons JGP, Schreiber RD, Perrin LH, Cooper NR, Muller-Eberhard HJ, Oldstone MBA (1979) Lysis of measles virus-infected cells by the purified cytolytic alternative complement pathway and antibody. J Exp Med 150: 445–454
Sissons JGP, Oldstone MBA, Schreiber RD (1980) Antibody-independent activation of the alternative complement pathway by measles virus-infected cells. Proc Natl Acad Sci USA 77: 559–562
Townsend ARM, Rothbard J, Gotch FM, Bahadur G, Wraith D, McMichael AJ (1986) The epitopes of influenza nucleoprotein recognized by cytotoxic T lymphocytes can be defined with short synthetic peptides. Cell 44: 959–968
van Bleek GM, Nathenson SG (1990) Isolation of an endogenously processed immunodominant viral peptide from the class I H-$2K^b$ molecule. Nature 348: 213–216

Walker RD, Plata F (1990) Cytotoxic lymphocytes against HIV. AIDS 4: 177–184
Whitton JL, Oldstone MBA (1989) Class I MHC can present an endogenous peptide to cytotoxic T lymphocytes. J Exp Med 170: 1033–1038
Whitton JL, Tishon A, Lewicki H, Gebhard J, Cook T, Salvato M, Joly E, Oldstone MBA (1989) Molecular analyses of a five amino acid cytotoxic T lymphocyte (CTL) epitope: an immunodominant region which induces nonreciprocal CTL cross-reactivity. J Virol 63: 4303–4310
Yanagi Y, Chan A, Chin B, Minden M, Mak TW (1985) Analyses of cDNA clones specific for human T cells and the alpha and beta chains, of the T-cell receptor heterodimer from a human T-cell line. Proc Natl Acad Sci USA 82: 3430–3434
Zinkernagel RM, Althage A (1977) Antiviral protection by virus-immune cytotoxic T cells: infected target cells are lysed before infectious virus progeny is assembled. J Exp Med 145: 644–651
Zinkernagel RM, Doherty PC (1974) Restriction of in vitro T cell-mediated cytotoxicity in lymphocytic choriomeningitis within a syngeneic or semiallogeneic system. Nature 248: 701–702

Therapeutic Reconstitution of Human Viral Immunity by Adoptive Transfer of Cytotoxic T Lymphocyte Clones

S. R. RIDDELL and P. D. GREENBERG

1 Introduction

The fundamental role of a competent host T lymphocyte response for promoting the resolution of acute viral infections is supported by the severe and progressive infections observed in individuals with primary or acquired deficiencies of T cell function (ROSEN et al. 1984a, b; REDFIELD et al. 1987;

Fred Hutchinson Cancer Research Center and the Departments of Medicine and Immunology, University of Washington, Seattle, WA 98104-2092, USA

Current Topics in Microbiology and Immunology, Vol. 189

ENGLUND et al. 1988; LJUNGMAN et al. 1989). The critical immune response is comprised of $CD8^+$ cytotoxic T cells (CTL) and $CD4^+$ helper T cells (Th), which recognize degraded fragments of proteins derived from pathogens and presented in association with class I and class II major histocompatibility complex (MHC) molecules, respectively (BRACIALE et al. 1987; YEWDELL and BENNINK 1990). $CD8^+$ CTL function to eradicate viral infection by the direct lysis of infected cells and by the secretion of cytokines such as γ-interferon. The predominant function of $CD4^+$ Th in the host response to infection is the production of cytokines, which in addition to direct antiviral effects facilitate the antiviral activities of macrophages, B cells, and other T cells including $CD8^+$ CTL.

For individuals with viral infections characterized by a latent and/or persistent phase after primary exposure, the maintenance of an MHC-restricted, host virus-specific T cell response is essential for controlling episodes of reactivation. There are striking clinical examples of the consequences of suppressing host immune competence including the fulminant infections that evolve in solid organ and bone marrow transplant (BMT) recipients due to reactivation of cytomegalovirus (CMV), varicella zoster virus (VZV), or herpes simplex virus (HSV), and the occurrence of Epstein Barr virus (EBV)-induced lymphoproliferative syndromes (PURTILO 1980; MEYERS et al. 1987; MARTIN et al. 1984; ZUTTER et al. 1988). The frequency and severity of many of these infections in immunocompromised hosts have been attenuated by the development of antiviral drugs but the requirement for long-term administration in chronically immunosuppressed hosts, the toxicity of the available antiviral agents, and the development of drug-resistant virus isolates has encouraged investigation of novel therapeutic approaches aimed at resolving the underlying immunodeficiency (SARAL et al. 1981; SCHMIDT et al. 1991; GOODRICH et al. 1991). One approach potentially applicable to human diseases and supported by a substantial body of experimental data in murine models is the use of adoptive immunotherapy with antigen-specific T cells to selectively reconstitute or augment deficient T cell responses.

2 Adoptive Immunotherapy of Viral Diseases in Murine Models

The principles for successful adoptive immunotherapy of human diseases have largely been established in experimental models for viral infections and tumors in mice (GREENBERG 1991). Studies in murine models have demonstrated that deficient protective T cell responses to viruses and tumors can be selectively reconstituted by the adoptive transfer of $CD4^+$ and $CD8^+$ antigen-specific T cells (LUKACHER et al. 1984; BYRNE and OLDSTONE 1984; LARSEN et al. 1984; GREENBERG et al. 1985; GREENBERG 1986). The demonstrations in mice that adoptive immunotherapy with $CD8^+$ CTL alone will protect hosts from infections with several viruses including murine CMV, influenza, HSV, and respiratory syncytial virus have provided convincing support for a decisive role of $CD8^+$ CTL responses in antiviral immunity (REDDEHASE et al. 1985; LUKACHER et al. 1984; LARSEN et al. 1984; CANNON et al. 1987). Moreover, the demonstrations that murine antigen-specific T cells isolated and cultured long-term in vitro prior to cell transfer retain efficacy in adoptive immunotherapy has established an important precedent for the treatment of human diseases in which in vitro expansion of effector T cells for therapeutic use is likely to be necessary (LUKACHER et al. 1984; BYRNE and OLDSTONE 1984; CHEEVER et al. 1986).

An extensively studied murine adoptive therapy model is the treatment of murine CMV infection in BALB/c mice. Although murine CMV is species specific and genetically disparate from human CMV, it shares many of the biological properties of the virus, including persistence as a latent infection in healthy hosts and reactivation in immunosuppressed hosts to cause a progressive infection culminating in pneumonitis and hematopoietic failure (GARDNER et al. 1974; JORDAN et al. 1977; SHANLEY et al. 1979; MUTTER et al. 1988).

Experiments in the murine CMV model have investigated the role of individual T cell subsets and T cells of selected antigen specificities in protection from lethal CMV infection. These studies have demonstrated that the adoptive transfer to immunosuppressed mice of T cells obtained from syngeneic animals previously immunized with murine CMV is protective against a lethal virus challenge. The transfer of purified $CD8^+$ T cells alone was sufficient to provide protection, whereas the transfer of purified $CD4^+$ T cells alone was not effective (REDDEHASE et al. 1985). The transfer of $CD8^+$ CTL also provided an antiviral effect in immunosuppressed mice with established murine CMV disease, demonstrating that the antiviral effects of $CD8^+$ CTL are not limited to the prophylactic setting (REDDEHASE et al. 1985).

Reconstitution of a protective response requires transfer of T cells with specificity for the appropriate viral antigens necessary to control and/or eliminate infection. Inoculation of BALB/c mice with murine CMV elicits an immunodominant $CD8^+$ CTL response specific for an epitope of the 89-kD a major immediate–early (IE) viral protein presented in association with single MHC allele, L^d (KOSZINOWSKI et al. 1987; REDDEHASE et al. 1989). Mice can be immunized with a recombinant vaccinia virus expressing the major IE protein, and the adoptive transfer of the $CD8^+$ T cells derived from such mice was found to be effective in protecting naive mice from challenge with a lethal inoculum of CMV, demonstrating that even a very limited repertoire of $CD8^+$ CTL can be sufficient for protective immunity (REDDEHASE et al. 1987a).

3 Cytomegalovirus Infection in Allogeneic Bone Marrow Transplant Recipients—A Model for Specific Adoptive Immunotherapy in Humans

The extension of adoptive immunotherapy from animal models to human diseases is predicated on the isolation and propagation of human antigen-specific T cells and on the ability of such T cells to persist and function in vivo following cell transfer. The first attempts at T cell transfer in humans evolved in the field of cancer immunotherapy (ROSENBERG et al. 1985). These studies have not specifically evaluated the use of MHC-restricted, antigen-specific T cells, but rather have utilized in therapy autologous polyclonal lymphocytes, derived from the peripheral blood or from a tumor biopsy, and activated in vitro with high concentrations of interleukin 2 (IL-2) (ROSENBERG et al. 1986, 1988). Up to 40% of patients with malignant melanoma have demonstrated antitumor responses, but only a small fraction of patients with tumors of other histologies have had measurable responses (ROSENBERG et al. 1988). Efforts to improve the response rate have been difficult, in part because the nature and specificity of the effector cells responsible for the observed antitumor effects have not been defined.

Infections that develop in immunodeficient allogeneic BMT recipients provide unique clinical situations in which the therapeutic and biologic effects of adoptive immunotherapy with T cells of defined specificity and function might be more readily evaluated. In this setting, T cells with the desired antigen specificity can potentially be isolated from the immunocompetent MHC-identical bone marrow donor, and the contribution of transferred effector T cells to host immunity can be readily evaluated in hosts deficient in the respective T cell response after ablative chemoradiotherapy. Our initial efforts in BMT recipients have been directed at developing adoptive immunotherapy as prophylaxis for human CMV infection. Reactivation of CMV occurs in 70%–80% of CMV-seropositive BMT recipients, and in the absence of antiviral therapy up to 50% of patients excreting virus will develop CMV pneumonia or enteritis (MEYERS et al. 1986, 1990). Combination antiviral therapy with ganciclovir and immunoglobulin has improved the survival for these patients. However, even with such therapy, 50% of patients developing CMV pneumonia still have a fatal outcome (REED et al. 1988; EMMANUEL et al. 1988). The median time after transplant to the development of clinically overt CMV disease is 50–60 days, presumably due to the slow replicative cycle of CMV, which allows adequate time to generate CMV-specific T cells in vitro from the bone marrow donor for adoptive transfer to the transplant recipient.

3.1 Contribution of $CD8^+$ and $CD4^+$ Major Histocompatibility Complex-Restricted T Cell Responses to Protection from Cytomegalovirus Disease

The development of effective adoptive immunotherapy for CMV infection after BMT would be facilitated by an understanding of the relative contributions of the $CD4^+$ and $CD8^+$ CMV-specific T cell responses to the control of viral reactivation. BMT recipients transplanted with bone marrow that has not intentionally been depleted of T cells receive with the marrow inoculum a small number of contaminating donor T lymphocytes of both the CD4 and CD8 phenotype. However, the absolute number of memory T cells specific for an individual pathogen is so low that recipients display a profound deficiency of antigen-specific T cell responses in the early post-transplant period. The tempo of immunologic recovery is variable in individual patients and influenced by several factors, including the preexisting immunity of the donor, the intensity of postgrafting immunosuppressive therapy, and the development of graft versus host disease (GVHD; WITHERSPOON et al. 1984; LUM 1987). Thus, the hypothesis that defects of CMV-specific T cell responses may contribute to the development of CMV disease can be directly tested by analyzing the temporal recovery of CMV-specific $CD4^+$ Th and $CD8^+$ CTL responses in individual BMT patients and correlating these responses with the occurrence of CMV disease.

Two studies have assessed $CD4^+$ CMV-specific Th responses of peripheral blood lymphocytes (PBL) obtained from BMT recipients and have demonstrated no correlation between the presence of $CD4^+$ CMV-specific Th and the clinical outcome of CMV infection (MEYERS et al. 1980; QUINNAN et al. 1982). In one of these studies, the ability of fresh PBL to directly lyse CMV-infected target cells in an in vitro assay over an extended duration was evaluated (QUINNAN et al. 1982). This analysis demonstrated that individuals with detectable cytotoxic responses to CMV-infected target cells were more likely to resolve a CMV infection. However, the cytolytic activity measured in this assay reflected both NK cell- and T cell-mediated lysis (QUINNAN et al. 1981). If NK cells were depleted prior to the assay, this method of directly examining T cells from PBL for cytolytic reactivity permitted detection of $CD8^+$ CTL responses only in patients with clinically overt CMV infection and was not sensitive enough to detect $CD8^+$ CTL in individuals without active infection (QUINNAN et al. 1982). Therefore, the specific contribution of endogenous reconstitution of $CD8^+$ CTL responses in preventing CMV infection or progression of subclinical reactivation to disease could only be inferentially determined.

The development of in vitro culture methods with greater sensitivity for detecting $CD8^+$ CMV-specific CTL responses in individuals without active infection has provided the opportunity for a more detailed examination of the role of $CD8^+$ CTL in the pathogenesis of CMV disease after BMT. Stimulation of PBL in vitro with autologous CMV-infected fibroblasts results in the selective expansion of $CD8^+$ CMV-specific CTL precursors which can be detected in a 4- to 6-h cytotoxicity assay (Borysiewicz et al. 1983; Schrier and Oldstone 1986). The exposure of target cells to γ-interferon prior to infection with CMV upregulates class I MHC gene expression and further increases the sensitivity of short-term cytolytic assays for detecting $CD8^+$ CMV-specific CTL (Laubscher et al. 1988).

Using such in vitro methods, the temporal recovery after allogeneic BMT of $CD8^+$ CMV-specific CTL responses was examined in samples of PBL obtained at 30, 60, and 90 days after transplant and the correlation between presence of these responses and occurrence of CMV disease determined. The initial analysis of 20 patients demonstrated that by day 90 after BMT, 50% of patients had recovered detectable $CD8^+$ CMV-specific CTL responses. Six of the ten patients without recovery of CTL had developed CMV pneumonia, whereas no cases of CMV pneumonia were observed in the ten patients with recovery of CTL responses (Reusser et al. 1991). This analysis of endogenous reconstitution of $CD8^+$ CMV-specific CTL responses has now been extended to include 58 patients, and the results confirm the initial observations—15 of the 58 patients developed CMV pneumonia, with none of the 15 having evidence of $CD8^+$ CMV-specific CTL responses in the assay prior to or concurrent with the onset of CMV disease (Table 1). Moreover, patients demonstrating recovery of $CD8^+$ CTL responses were protected from the subsequent development of CMV disease.

A simultaneous analysis of the recovery of $CD4^+$ CMV-specific Th responses in the 58 patients demonstrated that all patients with recovery of $CD8^+$ CMV-specific CTL had prior or concurrent recovery of $CD4^+$ CMV-specific Th responses. This would be expected, since recovery of $CD4^+$ Th is likely to be obligatory for the development of detectable $CD8^+$ CTL responses. Althouth a significant correlation between the recovery of $CD4^+$ Th

Table 1. The recovery of cytomegalovirus (CMV)-specific T cell responses after allogeneic bone marrow transplantation (BMT) correlates with protection from CMV pneumonia

	Occurrence of CMV pneumonia (*n*)	Total no. of patients
No recovery of CD4 or CD8	13*	26
Recovery of CD4 and CD8	0*	25
Recovery of CD4 alone	2	7
Recovery of CD8 alone	0	0

$CD8^+$ cytotoxic T lymphocyte and $CD4^+$ T-helper responses were evaluated in samples of peripheral blood lymphocytes obtained at days 30–40, 60–70, and 90–120 after BMT. The responses were assayed and scored as previously described (Reusser et al. 1991). The diagnosis of CMV pneumonia required the presence of CMV in tissue sections or bronchoalveolar lavage with compatible clinical signs and symtoms. Statistical analysis was performed using Fischer's exact test.
* $p < 0.01$

and protection from CMV disease was observed, two of seven patients with isolated recovery of weak but detectable $CD4^+$ Th responses developed CMV pneumonia, suggesting that this response alone is not sufficient to provide protection (Table 1). Thus, in this temporal analysis of endogenous immune reconstitution in allogeneic BMT recipients, the absence of $CD8^+$ CMV-specific CTL was found to be a major contributing factor to the development of CMV disease, suggesting that early immunologic reconstitution of this response by adoptive immunotherapy might be of significant therapeutic benefit.

3.2 In Vitro Isolation and Expansion of $CD8^+$ Cytomegalovirus Specific CTL Clones

The results from the analysis of endogenous reconstitution of CMV-specific T cell responses in BMT recipients focused our initial efforts on the development of in vitro techniques to isolate and expand $CD8^+$ class I MHC-restricted, CMV-specific CTL for potential use in adoptive immunotherapy. Although polyclonal T cell cultures enriched for $CD8^+$ CMV-specific CTL by several cycles of stimulation with autologous CMV-infected fibroblasts could be readily derived from CMV-seropositive bone marrow donors for use in therapy, this method would not likely provide a population of T cells with sole specificity for CMV antigens, and the adoptive transfer of such polyclonal CTL would carry the risk of inducing GVHD. This concern was reaffirmed by the observation that in some cultures of PBL stimulated with autologous CMV-infected fibroblasts, the CTL lines exhibited specificity for both CMV-infected autologous targets and uninfected allogeneic targets (RIDDELL et al. 1991a). To circumvent this problem, limiting dilution culture methods were developed to isolate individual $CD8^+$ CMV-specific CTL from cultures enriched for $CD8^+$ CMV-specific CTL activity (RIDDELL and GRENBERG 1990; RIDDELL et al. 1991b). The cloning efficiency of $CD8^+$ T cells plated with autologous CMV-infected fibroblasts as stimulator cells, γ-irradiated PBL, and EBV-transformed lymphoblastoid cell lines (LCL) as feeder cells, and recombinant IL-2 varied in different individuals from 2%–10% of plated T cells, with 30%–80% of clones isolated displaying class I MHC-restricted, CMV-specific cytolytic reactivity (RIDDELL et al. 1991b). The clonal origin of the T cells obtained in these cultures was confirmed by restriction endonuclease digestion of genomic DNA and analysis of T cell receptor gene rearrangements.

The number of T cells necessary to mediate a therapeutic effect following adoptive transfer in humans is not known, but estimates derived by extrapolation from animal models would suggest that between 10^8 and 10^{10} cells may be required for efficacy. As described above, $CD8^+$ CMV-specific CTL clones can be induced to proliferate in vitro with retention of MHC-restricted cytolytic function by intermittent stimulation with autologous CMV-infected

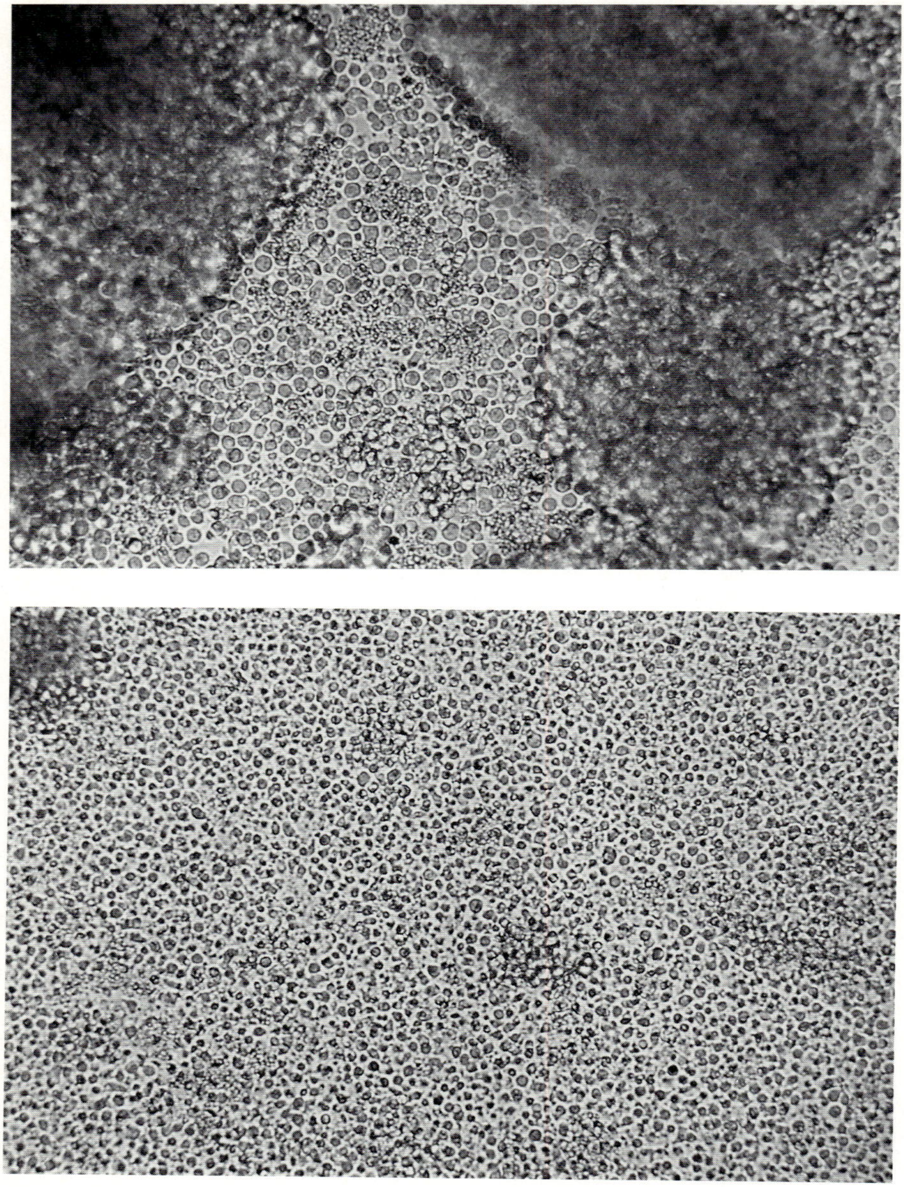

a

b

Fig. 1a, b. Morphology of a CD8$^+$ cytomegalovirus (CMV)-specific T cell clone during growth and resting phases after antigen stimulation. A CD8$^+$ CMV-specific T cell clone was stimulated with autologous CMV-infected fibroblasts at a responder to stimulator ratio of 20 and interleukin 2 (50 U/ml) was added to the culture at 2 and 4 days after stimulation. Panel a shows the T cells on day 5 after stimulation and illustrates the large, irregular morphology and clusters indicative and proliferating cells. Panel **b**, shows T cells of the same clone entering the resting phase 10 days after antigen stimulation, at which time the cells are small, round, and nonclustered fibroblasts and addition to the cultures of irradiated feeder cells and IL-2. After antigen stimulation and the addition of exogenous IL-2, T cells become large, form clusters, and undergo several days of proliferation before assuming a small, round morphology typical of resting lymphocytes prior to the next cycle of stimulation (Fig. 1). T cell clones grown in this fashion can be plated with irradiated feeder cells in the absence of antigen or IL-2 stimulation and will remain viable in vitro for long periods of time ($>$ 21 days; RIDDELL et al. 1991a). In contrast, T cells cultured in continously high concentrations of IL-2 without cyclical stimulation die rapidly when IL-2 is withdrawn (LENARDO 1991). Moreover, cycles of growth and rest, rather than continuous IL-2-driven expansion, may be important for retention of antigen specificity. The majority of $CD8^+$ CMV-specific T cell clones generated can be grown to 10^8 cells over 8 weeks and some can be expanded to more than 10^{10} cells 12–16 weeks after initial plating (RIDDELL et al. 1991a). To maintain large T cell numbers in these cultures, fibroblasts are grown in large flasks or roller bottles for use as stimulators, and donor PBL are obtained by leukapheresis and cryopreserved for use as feeder cells. Thus, these culture methods are adequate to provide the numerical expansion of human $CD8^+$ CMV-specific CTL clones predicted to be necessary for successful adoptive immunotherapy.

4 Evaluation of the Antigen Specificity of Protective $CD8^+$ Cytomegalovirus-Specific CTL

Despite the apparent importance of $CD8^+$ CTL responses in controlling infection with CMV, the role of specific viral proteins as target antigens for CTL has only recently been evaluated. In murine studies, the specificity of CTL responses to viruses is frequently biased towards a limited number of octa or nonapeptide epitopes derived from a relatively few immunodominant viral proteins and conforming to a binding motif specific for individual MHC molecules (VAN BLEEK and NATHENSON 1990). It is presumed that the immunodominant response detected in healthy individuals resistant to disease reflects the response most effective for preventing disease, although viral factors may also contribute to this limitation of the T cell repertoire (WOLD and GOODING 1991; GOODING 1992; POSAVAD and ROSENTHAL 1992). The use in adoptive immunotherapy of a small number of T cells clones, each specific for a single peptide, may impose a significant constraint on efficacy if the target epitopes are not representative of the immunodominant response or are subject to mutational variation in the pathogen. Thus, it became essential to elucidate whether particular CMV antigens were immunodominant in individuals with protective immunity and to determine the potential for variability of target antigens in different CMV strains.

4.1 Immunodominant Cytomegalovirus-Specific CD8$^+$ CTL Responses Recognize Structural Virion Proteins

The standard method to identify target antigens for CTL employs recombinant vectors to selectively express individual viral genes; however, this could not be efficiently applied to human CMV, as approximately 200 viral proteins may be encoded by the 230-kb CMV genome. Studies with influenza virus had suggested that antigens presented by class I MHC to CD8$^+$ CTL were predominately derived from the processing of newly synthesized viral proteins (BRACIALE et al. 1987). Since the transcription and translation of CMV genes after viral infection occurs sequentially (WALTHEN and STINSKI 1982), metabolic inhibitors could be used to selectively limit viral gene expression at distinct phases of the viral replicative cycle to identify the sets of genes encoding potential target antigens for CMV-specific CTL.

The first viral proteins synthesized after CMV infection are the products of the IE genes, which are required for the activation and subsequent expression of the early (E) genes (STINSKI et al. 1983). The late (L) genes are expressed following the onset of viral replication and encode the majority of the proteins comprising the virion structure (LANDINI and MICHELSON 1988). A sequential metabolic blockade of cells infected with CMV using cycloheximide, an inhibitor of protein synthesis, followed by actinomycin D, and inhibitor of RNA synthesis, will resı it in limiting the expression of viral proteins in infected cells to those encoded by the IE genes (REDDEHASE and KOSZINOWSKI 1984). Similarly, target cells will selectively express only IE and E proteins if infected with CMV in the presence of phosphonoformic acid (GRIFFITHS and GRUNDY 1987). Initial experiments using these approaches suggested that a high frequency of CD8$^+$ CTL from latently infected individuals recognized target cells expressing only IE, or IE and E gene products (BORYSIEWICZ et al. 1988; RIDDELL et al. 1991b). CD8$^+$ CMV-specific CTL clones were isolated to further define the specificity of these responses. Although the majority of CTL clones lysed target cells blocked with cycloheximide and actinomycin D, none of these clones lysed target cells infected with a vaccinia recombinant virus encoding the 72-kDa major IE protein (RIDDELL et al. 1991b). These results suggested that either minor IE proteins were the dominant target antigens for CTL or that structural virion proteins entering the cytoplasm after viral penetration were being processed and presented by the class I pathway prior to de novo endogenous synthesis.

To determine the potential role of structural CMV proteins entering with the virion as target antigens for CD8$^+$ CTL, autologous fibroblasts were pretreated with actinomycin D, infected with CMV, and assayed as targets for CTL in the continued presence of actinomycin D. Under these conditions, no transcription or translation products of the IE genes were detectable in infected cells, and thus the only viral antigens available for class I processing were those proteins entering with the virus (RIDDELL et al. 1991b). Remarkably, actinomycin D-blocked target cells were efficiently lysed in an MHC-

Table 2. Polyclonal CD8$^+$ cytomegalovirus (CMV)-specific cytotoxic T lymphocytes (CTL) recognize autologous CMV-infected cells in the absence of endogenous viral gene expression

Donor CTL	Target cell (% lysis)					
	Autologous			MHC mismatched		
	CMV	CMV/ActD	Mock	CMV	CMV/ActD	Mock
SS	46	37	1	8	6	2
CG	23	19	1	4	1	0
MR	64	57	1	12	3	8
BH	48	40	1	6	4	3
KD	66	64	2	4	1	1
DH	48	46	1	7	3	3
TM	45	48	6	5	2	2
CM	39	37	1	2	2	3
5937	42	43	3	2	3	3
6213	35	39	0	3	1	1

Polyclonal CD8$^+$ CTL lines were generated from peripheral blood lymphocytes obtained from CMV-seropositive individuals (designated by initials or a unique donor number as shown under the column "Donor CTL"). The CTL cultures were assayed 12–14 days after initiation for cytolytic reactivity against Cr^{51}-labeled autologous and major histocompatibility complex (MHC)-mismatched fibroblasts either mock infected (Mock), infected with CMV for 16 h (CMV), or infected with CMV for 4 h in the presence of the transcriptional inhibitor actinomycin D (CMV/ActD). Data is shown for an effector to target ratio of 10 or 20:1.

restricted manner by polyclonal CTL lines generated from ten consecutive CMV-seropositive individuals (Table 2). Moreover, the overwhelming majority of CD8$^+$ CMV-specific CTL clones isolated from these cultures also efficiently lysed target cells infected with CMV in the presence of actinomycin D. Conversely, the rare CTL clones we have isolated that are specific for the 72-kDa IE protein lyse autologous cells infected with the Vac/IE recombinant or with CMV, but as predicted do not lyse cells infected with CMV in the presence of actinomycin D (RIDDELL et al. 1991b). These results, which demonstrate an immunodominant role for introduced structural virion proteins as target antigens for the CD8$^+$ CMV-specific CTL response present in individuals with protective immunity, are consistent with recent results in experimental models of antigen processing demonstrating that only cytosolic location and not endogenous synthesis is required for proteins to be processed and presented by the class I pathway (MOORE et al. 1988; YEWDELL et al. 1988).

To gain insight into the variability in the epitopes recognized by CMV-specific CTL, we have analyzed the ability of polyclonal CTL lines and T cell clones to lyse target cells sharing only a single MHC allele with the donor. In the CMV-seropositive donors studied, the CTL response is predominantly restricted by one or at most two of the class I MHC molecules expressed by the individuals' cells with minor responses restricted by the other alleles

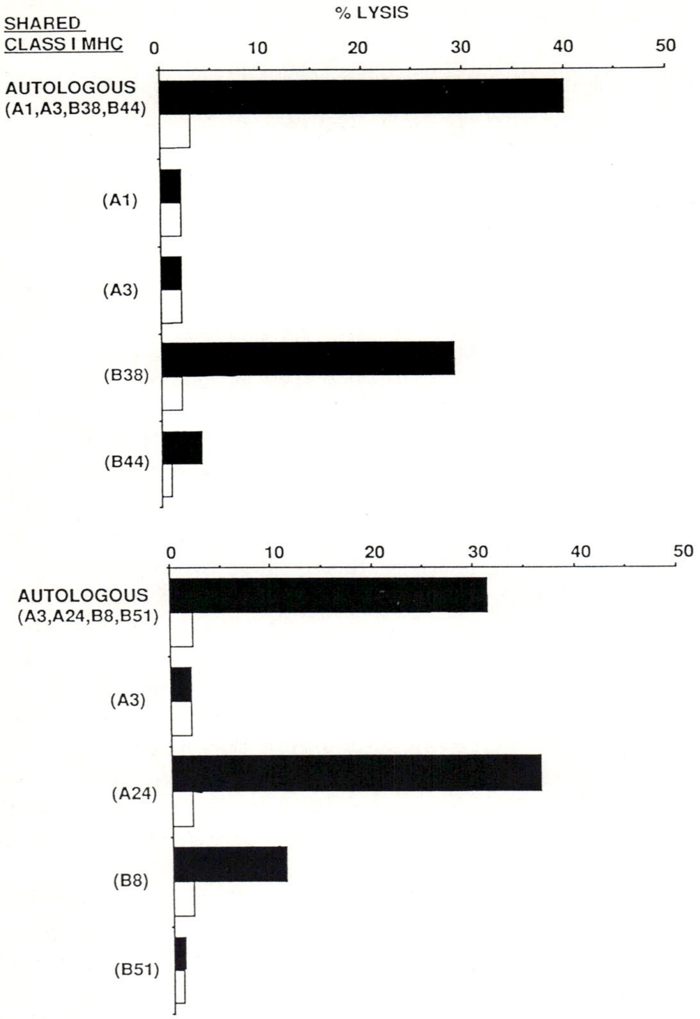

Fig. 2. Polyclonal $CD8^+$ cytomegalovirus (CMV)-specific cytotoxic T lymphocyte (CTL) responses in CMV-seropositive individuals are predominantly restricted by a single major histocompatibility complex (MHC) allele. Polyclonal CMV-specific CTL lines were generated by stimulation of peripheral blood lymphocytes with autologous CMV-infected fibroblasts as previously described (RIDDELL et al. 1991b). Cultures were restimulated once and assayed 12–14 days after initiation for recognition of autologous CMV-infected fibroblasts (■), autologous mock-infected fibroblasts (□), and allogeneic fibroblasts sharing only a single class I MHC molecule with the donor and either infected with CMV (■) or mock infected (□). The data is presented at an effector-to-target ratio of 10:1. The class I MHC phenotype of the donor target cell is given in *parentheses* under *Autologous* and the single class I MHC allele shared by allogeneic target cells is indicated in *parentheses*

(Fig. 2). Thus, despite the large number of potential target antigens encoded by CMV, the experimental evidence suggests that the host $CD8^+$ CTL response to CMV is predominantly focused on a small number of epitopes derived from proteins comprising the virion structure. This immunodominant CTL response, capable of recognizing infected cells prior to the onset of viral replication, may be particularly beneficial to the host by controlling viral spread after episodes of reactivation.

4.2 $CD8^+$ CTL Recognition of Cytomegalovirus Proteins in the Absence of Viral Gene Expression Is Selective

At least 20 proteins comprise the virion structure and represent candidate target antigens for the immunodominant CTL response to CMV. The most abundant of these include the envelope proteins GB and GH, the matrix proteins pp 65 and pp 150, and the 150-kDa major capsid protein (LANDINI and MICHELSON 1988; RASMUSSEN 1990; JAHN et al. 1987). The contribution of the envelope proteins GB and GH as target antigens for $CD8^+$ CMV-specific CTL clones representative of the immunodominant response has been evaluated using vaccinia GB and GH recombinant viruses to infect target cells and selectively express the GB or GH proteins. In these experiments, CTL clones that recognize GB- or GH-expressing target cells were not identified. Although GB was not the target antigen of the immunodominant CTL response, $CD8^+$ CTL clones specific for GB could be isolated from seropositive individuals by infection of autologous stimulator cells with Vac/GB to achieve selective expression of the GB protein (RIDDELL et al. 1991b). This was consistent with the results of a limiting dilution analysis in two CMV-seropositive individuals demonstrating a low precursor frequency of GB-specific CTL (BORYSIEWICZ et al. 1988).

The relative absence of CTL specific for GB or GH may reflect the propensity with which these proteins are accessable to the class I antigen processing and presentation pathway early after CMV infection. To determine how efficiently GB was processed for class I presentation compared with proteins recognized by the immunodominant CTL response, we examined the ability of $CD8^+$ GB-specific CTL to lyse cells infected with CMV at various stages of the replicative cycle. GB-specific CTL do not lyse cells infected with CMV in the presence of an actinomycin D blockade, but will lyse cells infected with CMV for longer than 24 h with no metabolic inhibitors (RIDDELL et al. 1991b). By contrast, cells infected with CMV are recognized by CTL representative of the immunodominant response at all stages of the replicative cycle. The failure to present GB envelope protein immediately after viral infection may be due in part to the intimate involvement of the viral envelope in fusion at the plasma membrane, resulting in limited amounts entering the cytosol and being available for presentation with class I MHC molecules. Consistent with CMV envelope proteins serving only a minor role as target antigens for CTL responses, efforts have been unsuccessful to isolate from several individuals CTL specific for a second envelope glycoprotein, GH, despite the use of a vac/GH recombinant virus to selectively express GH in stimulator cells.

The structural proteins comprising the matrix and capsid of CMV are not bound to the cell membrane, but rather enter the cytoplasm following viral fusion. The most abundant protein component of the matrix of CMV is a 65-kDa phosphoprotein (pp65; JAHN et al. 1987). CTL specific for pp65 were identified in the polyclonal CTL response elicited by in vitro stimulation with CMV-infected fibroblasts in the majority of individuals studied, and approximately 50% of the CMV-specific CTL clones isolated from one individual recognized autologous targets infected with Vac/pp65. Moreover, the pp65-specific CTL clones lysed cells infected with CMV in the presence of an actinomycin D blockade, demonstrating that pp65 can be directly processed and presented to $CD8^+$ CTL without the need for endogenous synthesis. The remainder of the $CD8^+$ CMV-specific CTL clones isolated from this individual recognized one or more structural proteins distinct from pp65 that could also be presented by cells infected in the presence of actinomycin D. Construction of recombinant vectors to express additional matrix or capsid proteins should elucidate the relative contributions of these proteins as target antigens for CTL responses to CMV.

4.3 Epitopes Recognized by $CD8^+$ CTL Are Conserved in Clinical Isolates of Cytomegalovirus

The majority of clinical isolates of CMV exhibit substantial genetic heterogeneity by restriction endonuclease digestion (HUANG et al. 1980; CHANDLER and McDOUGALL 1986). It is unknown whether this genetic variability confers differences in the pathogenesis of CMV infection. However, the recent identification of subgroups of CMV based on heterogeneity in the envelope glycoprotein B gene demonstrates that significant variation may occur in viral genes that encode target antigens for CTL (CHOU and DENNISON 1991). Individuals undergoing allogeneic BMT are potentially at risk for exposure to several genetically different strains of CMV, including the virus transferred with the donor bone marrow inoculum, the endogenous latent CMV strain, and CMV strains transmitted by blood products administered as supportive care after the BMT (CHOU 1986). Thus, it would be advantageous if the T cell clones used in adoptive therapy recognized highly conserved epitopes present in otherwise genetically distinct clinical isolates. Experiments were performed in five individuals in whom the immunodominant response was restricted by different class I MHC molecules to analyze the recognition by CTL of target cells infected either with laboratory strains AD169 or Towne, or with five genetically distinct CMV isolates recently derived from BMT recipients. The epitopes recognized by the $CD8^+$ CTL clones representative of the immunodominant response in these individuals were conserved in all the CMV strains tested, suggesting that the genetic heterogeneity of the virus is unlikely to present a significant obstacle to successful adoptive immunotherapy (GREENBERG et al. 1991).

5 Restoration of CD8$^+$ CTL Responses in Bone Marrow Transplant Recipients by Adoptive Transfer of CTL Clones

The results of our studies investigating the immunobiology of CMV infection after allogeneic BMT supported a crucial role for CD8$^+$ CTL responses in preventing CMV disease and suggested that the adoptive transfer of T cell clones to reconstitute immunity warranted investigation as a prophylactic or therapeutic strategy for CMV infection in BMT recipients. The objectives of our initial adoptive therapy experiments in allogeneic BMT recipients were to determine whether the adoptive transfer of CD8$^+$ CMV-specific CTL clones could be accomplished safely and if transferred T cells could persist and function in vivo. T cell infusions were administered to the patients beginning 28–35 days after BMT. This time point was selected for several reasons. First, differentiation of complications due to the transplant-conditioning regimen, which predominantly occur in the first 21 days after BMT, could be more easily distinguished from toxicities that might occur as a consequence of T cell therapy. Second, our prior studies demonstrated that only 25%–35% of patients have endogenous recovery of CD8$^+$ CMV-specific CTL responses at days 30–40 after transplant (REUSSER et al. 1991; LI and RIDDELL, unpublished data); thus, it should be possible to measure a contribution of transferred CTL in the majority of patients. Third, reconstitution of protective responses would begin before the highest period of infectious risk, since less than 2% of CMV-seropositive BMT patients will develop CMV disease prior to day 35.

Seven CMV-seropositive patients receiving a BMT from CMV-seropositive, HLA-identical related donors have been entered into this study. CD8$^+$ CMV-specific T cell clones were successfully generated from each of the seven bone marrow donors and propagated to cell numbers sufficient for therapy in all patients. Two patients did not receive T cell infusions—one died of bacterial sepsis prior to day 35 and one patient developed severe pulmonary toxicity related to the BMT-conditioning regimen and did not meet the eligibility criteria for T cell therapy at day 35 after BMT. Five patients have received all of the four scheduled T cell infusions.

5.1 Generation and Characterization of CTL Clones for Immunotherapy

Fibroblast lines for use as stimulator and target cells were established from skin biopsies obtained from the MHC-identical bone marrow donors approximately 4 weeks prior to the scheduled BMT. Polyclonal $CD8^+$ CMV-specific CTL were generated by stimulating PBL obtained from the bone marrow donor with autologous CMV-infected fibrolasts, and T cell clones were isolated from the polyclonal CTL cultures by limiting dilution cloning. Screening assays were used to identify $CD3^+$ $CD8^+$ $CD4^-$ T cell clones which lysed autologous CMV-infected fibroblasts in a class I MHC-restricted manner, but failed to lyse uninfected fibroblasts. Clones were also assayed for lysis of fibroblasts and EBV-transformed B cells derived from the BMT recipient to identify and exclude from use in therapy any CTL with cross-reactivity for recipient minor histocompatibility antigens. CTL clones meeting these selection criteria were propagated to large numbers over 5–12 weeks. Prior to administration to the BMT recipient, all clones were assessed for sterility and mycoplasma contamination.

5.2 Clinical Monitoring of Patients Receiving Adoptive Immunotherapy with $CD8^+$ CTL

Significant toxicities including fevers, fluid retention, renal failure, and pulmonary edema have been observed in cancer patients receiving infusions of lymphokine-activated killer (LAK) cells or tumor-infiltrating lymphocytes (TIL) in conjunction with IL-2 (ROSENBERG et al. 1986, 1988). The toxicities observed are identical to those seen when IL-2 is administered alone, making assessment of the contribution of the cellular component of this therapy difficult. IL-2-activated killer cells localize in the lungs immediately after infusion, due in part to activation-induced adherence to endothelium, and therefore might contribute to the pulmonary compromise observed in these patients (FISHER et al. 1989). Since, in the adoptive transfer studies in CMV, IL-2 is not administered with or following the transfer of $CD8^+$ CMV-specific CTL, the magnitude of the toxicities observed in the LAK and TIL trials would not be expected.

To minimize any potential side effects of T cell therapy and to provide insight into the number of T cells required to reconstitute detectable responses, one infusion of $CD8^+$ CTL was administered each week for 4 consecutive weeks. Providing no toxicity was encountered, the dose of T cells administered was escalated according to the following schedule: first infusion—3.3×10^7 T cells/m²; second infusion—1×10^8 cells/m²; third infusion—3.3×10^8 T cells/m²; and fourth infusion—1×10^9 T cells/m². Although up to ten individual T cell clones were administered with each infusion to achieve the desired cell dose, many doses, including the highest dose, could be achieved with one or a composite of two clones.

All patients were monitored for alterations in vital signs and oxygen saturation during, and for 2 h after completion of, the T cell infusions. There were no significant changes in vital signs or oximetry recordings during this monitoring period with any of the 20 T cell infusions. One patient developed a fever 3 h following the third infusion, but defervesced without antimicrobial therapy. Given that fevers are relatively common after BMT, it was not evident that this temperature elevation was related to T cell therapy. After the initial three patients had completed therapy, subsequent T cell infusions were administered to them in the outpatient department. In these initial patients, no significant immediate toxicities due to T cell transfer were apparent.

Although the T cells used in adoptive transfer are of donor origin, the selection of clones with class I MHC-restricted, CMV-specific cytolytic reactivity should preclude a significant risk of inducing or worsening GVHD. All patients on this study were evaluated for changes in GVHD status on a daily basis while inpatients and twice weekly as outpatients. None of the five patients had any evidence of progression of GVHD during the course of T cell therapy.

5.3 Immunologic Monitoring of Cytomegalovirus-Specific T Cell Responses After Adoptive Transfer of CD8$^+$ CTL

The reconstitution of T cell immunity to CMV in treated patients was evaluated by assaying PBL obtained before initiating T cell therapy and 48 h after each infusion for CMV-specific CD8$^+$ CTL and CD4$^+$ Th responses. Consistent with the severe immunodeficiency observed at this time after allogeneic BMT, three of the five patients had no detectable CD8$^+$ CMV-specific CTL or CD4$^+$ Th responses before T cell transfer, and two patients had weak but detectable responses. In each of the three deficient patients, weak CD8$^+$ class I MHC-restricted CTL responses were detected 48 h after the first T cell infusion and the responses were augmented after each subsequent infusion (RIDDELL et al. 1992a). The CTL responses persisted between infusions, as reflected by assaying samples obtained just prior to the third infusion, implying that the increasing responses observed after each infusion reflected both persistent and recently infused CTL (Table 3). In these three patients, CD8$^+$ CTL responses were achieved by the third infusion that were equivalent in magnitude to those present in the healthy bone marrow donor, despite no evidence of recovery of CD4$^+$ CMV-specific Th responses during the 4 weeks of therapy. It must be emphasized that in 58 patients studied for endogenous reconstitution of CMV-specific T cell responses after allogeneic BMT, the recovery of CD4$^+$ Th responses was

Table 3. Reconstitution of CD8^{+} cytomegalovirus (CMV)-specific cytotoxic T lymphocyte (CTL) responses in allogeneic bone marrow transplant (BMT) recipients by the adoptive transfer of T cell clones

T cell infusion	Patient no. 6032				Patient no. 6131 Target cell (% lysis)				Patient no. 6025			
	Autologous		Allogeneic		Autologous		Allogeneic		Autologous		Allogeneic	
	CMV	Mock	CMV	Mock	CMV	Mock	CMV	Mock	CMV	Mock	CMV	Mock
Pre #1	2	1	ND	ND	7	6	1	0	4	0	0	1
Post #1	17	4	ND	ND	23	2	1	0	14	0	1	0
Post #2	43	5	6	12	48	3	6	3	12	1	1	1
Pre #3	37	5	3	6	20	2	0	1	12	1	1	1
Post #3	53	8	9	11	33	3	0	1	48	6	2	2
Post #4	83	10	4	5	38	2	2	4	55	6	0	0
Bone marrow donor	53	1	2	2	37	5	4	1	29	1	1	3

CD8^{+} CMV-specific CTL responses in BMT recipients and their respective bone marrow donors were evaluated in peripheral blood lymphocytes by a CTL generation assay (Riddell et al. 1991b). Peripheral blood lymphocytes were obtained from the recipients immediately before initiating T cell therapy (pre #1), 2 days after each T cell infusion (post #1 to post #4), and just before the third infusion (pre #3). CTL generation cultures were assayed in a 5-h chromium-release assay for class I major histocompatibility complex (MHC)-restricted, CMV-specific cytolytic reactivity 14 days after initiation. Target cells were autologous and allogeneic class I MHC-mismatched fibroblasts either infected with CMV or mock infected. The data are shown for an effector-to-target ratio (E/T) of 20:1 for patient 6032 and the respective bone marrow donor, and at an E/T of 10:1 for patients 6131, 6025, and their respective donors.

obligatory for the endogenous recovery of CD8^{+} CTL (Table 1). Thus, these functional studies provide compelling evidence for the efficacy of adoptive transfer of CTL clones in restoring CD8^{+} CMV-specific CTL responses in immunodeficient hosts. Two of the five patients had weak but detectable CD8^{+} CMV-specific CTL responses and readily detectable CD4^{-} Th responses at the onset of therapy. Substantial augmentation of CD8^{+} CTL responses was observed 48 h after the T cell infusions, consistent with the successful transfer of additional effector T cells, although the possibility of endogenous recovery of CD8^{+} CTL responses could not be excluded in these two patients.

All patients in this study received as prophylaxis for GVHD immunosuppressive therapy consisting of methotrexate and cyclosporine (three patients), methotrexate alone (two patients), or cyclosporine and prednisone (two patients). Since post-transplant immunosuppressive therapy contributes to the prolonged period of impaired immune competence observed after allogeneic BMT, the therapeutic success of adoptive T cell transfer is in part dependent on the ability of transferred T cells to persist in vivo. Observation of the three patients who were deficient in CD8^{+} CTL and CD4^{+} Th responses prior to adoptive immunotherapy proved to be particularly informative in the evaluation of long-term persistence of transferred CTL. PBL were obtained at 2 and 4 weeks after the completion of T cell therapy and assayed for CMV-specific CD8^{+} CTL and CD4^{+} Th responses. After 2 weeks, the magnitude of the CD8^{+} CTL responses in the three patients had declined, but was still evident, despite the continued absence of a CD4^{+} Th response, consistent with a limited survival of a component of the transferred CTL and/or with migration of the infused CTL from the peripheral blood to tissue sites. At 4 weeks, CD8^{+} CMV-specific CTL responses remained easily detectable in the three patients, but one of the three patients had weak endogenous recovery of a CD4^{+} CMV-specific Th response, making it possible that endogenous recovery of CD8^{+} CTL was also contributing to the responses detected (Tables 4, 5). In the other two patients, no recovery of CD4^{+} CMW-specific Th responses was detected, demonstrating that adoptively transferred CTL clones persist in vivo for at least 1 month after transfer (RIDDELL et al. 1992a). One of the five patients received only a small number of distinct T cell clones, and the unique rearrangements of the T cell antigen receptor genes are being used to identify transferred T cells persisting in the host at later times after T cell therapy. Determining the survival of individual clones in larger numbers of patients would be facilitated by the introduction of a marker gene by gene transfer techniques as has been applied to studies of adoptive immunotherapy in cancer patients (ROSENBERG et al. 1990). Such studies to determine the duration of in vivo survival of adoptively transferred T cell clones and the potential survival advantage of individual clones should be helpful in determining the optimal cell doses and infusion intervals for future studies in CMV and other diseases.

Table 4. Persistence of adoptively transferred CD8$^+$ cytomegalovirus (CMV)-specific cytotoxic T lymphocyte (CTL) responses after adoptive immunotherapy with CD8$^+$ T cell clones

	Patient no. 6032				Patient no. 6131 Target cell (% lysis)				Patient no. 6025			
	Autologous		Allogeneic		Autologous		Allogeneic		Autologous		Allogeneic	
	CMV	Mock	CMV	Mock	CMV	Mock	CMV	Mock	CMV	Mock	CMV	Mock
Post #4	83	10	4	5	38	2	2	4	55	6	0	0
2 weeks post #4	31	2	1	1	30	3	3	2	31	3	2	2
4 weeks post #4	39	5	4	5	31	4	5	4	41	7	2	1

CD8$^+$ CMV-specific CTL responses of PBL obtained from recipients 0, 2, and 4 weeks after completion of the fourth (#4) T cell infusion were evaluated using the CTL generation assay. Target cells include autologous fibroblasts either infected with CMV or mock infected, and HLA-mismatched (allogeneic) fibroblasts either infected with CMV or mock infected. Data are shown for an effector-to-target ratio (E/T) of 20:1 for patient 6032 and an E/T of 10:1 for patients 6131 and 6025.

Table 5. CD4$^+$ cytomegalovirus (CMV)-specific T-helper (Th) responses in recipients of adoptive immunotherapy with CD8$^+$ CMV-specific cytotoxic T lymphocyte

	Patient no. 6032			Patient no. 6131			Patient no. 6025		
	Control	CMV Ag	PHA	Control	CMV Ag	PHA	Control	CMV Ag	PHA
Post #4	1412	560	9951	179	143	22882	466	721	4964
2 weeks post #4	639	861	11255	142	493	41976	121	141	4511
4 weeks post #4	1409	1413	12376	408	1005	56590	201	292	5499
Bone marrow donor	2099	42664	145801	1319	20796	445524	1502	9624	160964

CD4$^+$ CMV-specific Th responses of peripheral blood lymphocytes (PBL) obtained from recipients 0, 2, and 4 weeks after completion of the fourth (#4) T cell infusion were evaluated in a 96-h proliferation assay. Triplicate cultures of 2×10^5 PBL per well were plated in 96 round-bottomed plates with a control antigen preparation derived by glycine extraction of mock-infected fibroblasts, a CMV antigen (CMV Ag) preparation derived by glycine extraction of CMV-infected fibroblasts, and with phytohemagglutinin (PHA, 10 μg/ml). Wells were pulsed with 2 μCi of tritiated thymidine for the final 18 h of the 96-h incubation. The results are shown as the mean counts per minute (cpm) of triplicate wells.

5.4 Virologic Monitoring of Patients Receiving Adoptive Immunotherapy with CD8$^+$ CTL

Reactivation of CMV as defined by a positive shell vial or conventional culture from throat, urine, or blood occurs in 70%–80% of CMV-seropositive individuals receiving an allogeneic BMT from a CMV-seropositive bone marrow donor (MEYERS et al. 1986). CMV viremia, which occurs in 40%–50% of patients, is highly predictive of the subsequent development of CMV disease and is thought to indicate substantial viral replication at sites of reactivation (MEYERS et al. 1990). All patients entered into this study had weekly monitoring of throat, urine, and blood by shell vial assays and culture for CMV to determine the potential antiviral effects of T cell transfer. The initial results of these virologic studies in the five patients are encouraging, although the number of patients treated with adoptive immunotherapy thus far is insufficient for definitive evaluation of the efficacy of reconstituting CD8$^+$ CTL responses in providing protection from CMV disease. None of the five patients treated with T cell infusions have developed CMV viremia or disease. One patient excreted CMV from the throat prior to receiving T cell therapy, and the cultures became negative in the week after the first T cell infusion and remained negative throughout the remainder of the post-BMT course. Three patients had isolated urinary excretion of CMV on a single occasion, as detected by shell vial assay in two and both shell vial assay and culture in one patient. Since normal immunocompetent individuals may excrete virus in the urine for prolonged periods of time after primary CMV infection despite clearance of viremia, the significance of isolated urinary excretion of CMV in these BMT patients is unclear, but may imply that additional host defense mechanisms other than CD8$^+$ CTL are necessary to maintain virus reactivation below a detectable level. The follow-up of a larger cohort of patients receiving adoptive immunotherapy to reconstitute CD8$^+$ CMV-specific CTL responses will assist in precisely defining the antiviral effects. However, the lack of toxicity and the absence of CMV disease observed in the first five patients suggests further investigation of this approach is warranted.

6 Conclusions

The use of adoptive immunotherapy to reconstitute or augment deficient antigen-specific T cell responses as treatment for pathogens and tumors in humans has long been an appealing concept. Results in animal models have established the principles of T cell transfer and validated the therapeutic promise of this approach for both malignant and infectious diseases. The initial investigations of adoptive immunotherapy in human malignant disease have been dissappointing and hindered by the difficulties in reliably isolating and propagating human T cells with MHC-restricted specificity for human tumors. Studies of the adoptive transfer of $CD8^+$ CMV-specific CTL in immunodeficient allogeneic BMT recipients now provide the first evidence in humans that adoptive transfer of T cell clones of defined specificity can be used successfully to restore antigen-specific T cell responses.

Adoptive immunotherapy with $CD8^+$ CMV-specific CTL resulted in no evident toxicities with single infusions of up to 2.2×10^9 T cells and total cell doses of 3.2×10^9. The successful transfer of CMV-specific CTL responses can be detected in the recipients' PBL with small cell doses ($3.3 \times 10^7/m^2$), and the responses measured after the transfer of 3.3×10^8 T cells/m^2 are equivalent in magnitude to those observed in healthy immunocompetent CMV-seropositive individuals. It is possible that the coadministration of $CD4^+$ CMV-specific Th or the administration of exogenous IL-2 would result in the amplification in vivo of lower doses of $CD8^+$ CTL, as has been demonstrated in murine adoptive transfer models (CHEEVER et al. 1982; REDDEHASE et al. 1987b). However, even in the absence of $CD4^+$ Th or IL-2, the data in these patients demonstrated that transferred $CD8^+$ CTL responses capable of functioning as memory T cells persist in the peripheral blood in vivo for at least 1 month. The precise duration of in vivo persistence will require analysis in patients receiving a small number of individual clones in order to facilitate the use of molecular analysis to determine the presence of CTL in PBL obtained at long intervals after therapy.

Allogeneic BMT recipients who recover endogenous $CD8^+$ CMV-specific CTL responses are protected from the development of CMV disease (REUSSER et al. 1991; LI and RIDDELL, unpublished data). Consistent with this, none of the patients who had $CD8^+$ CTL responses reconstituted by adoptive transfer developed CMV disease. However, additional studies of this approach for CMV are required to fully elucidate the biologic and therapeutic consequences of restoring immunity by the adoptive transfer of antigen-specific T cells. Our initial studies of the transfer of $CD8^+$ CTL clones were restricted to patients undergoing allogeneic BMT, but have implications for the use of T cell transfer as a means of diminishing the frequency and severity of CMV disease in solid organ transplant recipients and in patients with acquired immunodeficiency syndrome (AIDS). The infusion of autologous expanded CMV-specific CTL may be particularly useful for AIDS patients with CMV infections, because antiviral drug therapy is often poorly tolerated in this patient population (JACOBSEN and MILLS 1988). The treatment of AIDS patients with active CMV viremia or tissue infection should provide additional insights into the therapeutic potential of $CD8^+$ CTL. Moreover, the use of CTL clones specific for HIV gene products might also be of therapeutic value for HIV infection, particularly if gene transfer approaches to modify the potential for CTL clones to survive and function in vivo prove to be successful (RIDDELL et al. 1992b).

The studies of adoptive T cell transfer for human viral infections also have implications for the treatment of human malignancies. The identification of antigens expressed by human melanoma and recognized by CD8+ CTL suggest that application of specific immunotherapy for this human tumor may soon be possible (VAN DER BRUGGEN et al. 1991). Moreover, the provocative experiments in murine models demonstrating that immunization of mice with syngeneic tumor cells genetically modified to express selected cytokine genes will elicit host CD4+ and CD8+ tumor-specific T cell responses suggests a novel strategy for inducing T cells reactive with tumor antigens expressed by tumors of diverse histologies (FEARON et al. 1990; GOLUMBEK et al. 1991). If such an approach is effective in humans, the feasibility of isolating T cells in vitro with MHC-restricted tumor specificity for use in adoptive immunotherapy will be greatly improved.

Acknowledgements. Supported by grant CA18029 from the National Cancer Institute and a grant from the Cancer Research Institute (SRR). Stanley R. Riddell is a Special Fellow of the Leukemia Society of America. We wish to thank William Britt, University of Alabama (Birmingham), for providing the vaccinia/CMV recombinant viruses.

References

Borysiewicz LK, Morris S, Page J, Sissons JGP (1983) Human cytomegalovirus-specific cytotoxic T lymphocytes: requirements for in vitro generation and specificity. Eur J Immunol 13: 804–809

Borysiewicz LK, Graham S, Hickling JK, Sissons JGP (1988a) Precursor frequency and stage specificity of human cytomegalovirus-specific cytotoxic T cells. Eur J Immunol 18: 269–275

Borysiewicz LK, Hickling JK, Graham S, Sinclair J, Cranage MP, Smith GL, Sissons JGP (1988b) Human cytomegalovirus-specific cytotoxic T cells: relative frequency of stage-specific CTL recognizing the 72 KD immediate early protein and glycoprotein B expressed by recombinant vaccina viruses. J Exp Med 168: 919–932

Braciale TJ, Morrison LA, Sweetser MJ, Sambrook J, Gething MJ, Braciale VL (1987) Antigen presentation pathways to class I and class II MHC restricted T lymphocytes. Immunol Rev 98: 94–114

Byrne JA, Oldstone MBA (1984) Biology of cloned cytotoxic T lymphocytes specific for lymphocytic choriomeningitis virus: clearance of virus in vivo. J Virol 51: 682–686

Cannon MD, Stott EJ, Taylor G, Askonsas BA (1987) Clearance of persistent respiratory syncytial virus infections in immunodeficient mice following transfer of primed T cells. Immunology 62: 133–139

Chandler SH, McDougall JK (1986) Comparison of restriction site polymorphisms among clinical isolates and laboratory strains of human cytomegalovirus. J Gen Virol 67: 2179–2192

Cheever MA, Greenberg PD, Fefer A, Gillis S (1982) Augmentation of the antitumor therapeutic efficacy of long-term cultured T lymphocytes by in vivo administration of purified interleukin 2. J Exp Med 155: 968–980

Cheever MA, Thompson DB, Klarnet JP, Greenberg PD (1986) Antigen-driven long-term cultured T cells proliferate in vivo, distribute widely, mediate specific tumor therapy and persist long term as functional memory T cells. J Exp Med 163: 1100–1112

Chou SW (1986) Acquisition of donor strains of CMV by renal transplant patients. N Engl J Med 314: 1418–1423

Chou SW, Dennison DM (1991) Analysis of interstrain variation in cytomegalovirus glycoprotein B sequences encoding neutralization-related epitopes. J Infect Dis 163: 1229–1234

Emmanuel D, Cunningham I, Joles-Elysec K, Brochstein JA, Kernan NA, Laver J, Stover D, White DA, Fels A, Polsky G, Castro-Malaspina H, Peppard JR, Bartus P, Hammerling U, O'Reilly RJ (1988) Cytomegalovirus pneumonia after bone marrow transplantation successfully treated with the combination of ganciclovir and high dose intravenous immune globulin. Ann Intern Med 109: 777–782

Englund JA, Sullivan CJ, Jordan MC (1988) Respiratory syncytical virus infection in immunocompromised adults. Ann Intern Med 109: 203–208
Erice A, Chou SW, Biron K, Stanat SC, Balfour HH, Jordan MC (1989) Progressive disease due to ganciclovir—resistant cytomegalovirus in immunocompromised patients. N Engl J Med 320: 289–293
Fearon ER, Pardoll DM, Itaya T, Golumbek P, Levitsky HI, Simons JW, Karasuyama H, Vogelstein B, Frost P (1990) Interleukin 2 production by tumor cells bypasses T helper function in the generation of an antitumor response. Cell 60: 397–403
Fisher B, Packard BS, Reed EF, Caraquillo JA, Carter CS, Topalian SL, Yang SC, Yolles P, Larson JM, Rosenberg SA (1989) Tumor localization of adoptively transferred indium-111 labeled tumor infiltrating lymphocytes in patients with metastatic melanoma. J Clin Oncol 7: 250–261
Gardner MB, Officer JE, Parker J, Estes JD, Rongey RW (1974) Induction of disseminated virulent cytomegalovirus infection by immunosuppression of naturally chronically infected wild mice. Infect Immun 10: 966–969
Golumbek PT, Lazenby AJ, Levitsky HI, Jaffee LM, Karasuyama H, Baker M, Pardoll DM (1991) Treatment of established renal cancer by tumor cells engineered to secrete interleukin-4. Science 254: 713–716
Gooding LR (1992) Virus proteins that counteract host immune defenses. Cell 71: 5–7
Goodrich JM, Mori M, Gleaves CA, DuMond C, Cays M, Ebeling DF, Buhles WC, deArmond B, Meyers JD (1991) Early treatment with ganciclovir to prevent cytomegalovirus disease after allogeneic bone marrow transplantation. N Engl J Med 325: 1601–1607
Greenberg PD (1986) Therapy of murine leukemia with cyclophosphamide and immune Lyt 2^+ cells: cytolytic T cells can mediate eradication of disseminated leukemia. J Immunol 136: 1917–1922
Greenberg PD (1991) Adoptive T cell therapy of tumors: mechanisms operative in the recognition and elimination of tumor cells. Adv Immunol 49: 281–335
Greenberg PD, Kern DE, Cheever MA (1985) Therapy of disseminated murine leukemia with cyclophosphamide and immune Lyt-1^+2^- T cells: tumor eradication does not require participation of cytotoxic T cells. J Exp Med 161: 1122–1135
Greenberg PD, Reusser P, Goodrich JM, Riddell SR (1991) Development of a treatment regimen for human cytomegalovirus (CMV) infection in bone marrow transplantation recipients by adoptive transfer of donor-derived CMV-specific T cell clones expanded *in vitro*. Ann NY Acad Sci 636: 184–195
Guillaume JC, Sasaz P, Wechsler J, Lescs MD, Rouzeau JC (1991) Vaccinia from recombinant virus expressing HIV genes. Lancet 337: 1034–1035
Griffiths PD, Grundy JC (1987) Molecular biology and immunology of cytomegalovirus. Biochem J 241: 313–324
Huang E-S, Alford CA, Reynolds DW, Stagno S, Pass RF (1980) Molecular epidemiology of cytomegalovirus infections in women and their infants. N Engl J Med 303: 958–962
Jacobson MA, Mills J (1988) Serious cytomegalovirus disease in the acquired immunodeficiency syndrome (AIDS). Ann Intern Med 108: 585–594
Jahn G, Scholl B-C, Traupe B, Fleckenstein B (1987) The two major structural phosphoproteins (pp65 and pp150) of human cytomegalovirus and their antigenic properties. J Gen Virol 68: 1327–1337
Jordan MC, Shanley JD, Stevens JG (1977) Immunosuppression reactivates and disseminates latent murine cytomegalovirus. J Gen Virol 37: 419–423
Koszinowski UH, Keil GM, Schwarz H, Schickedanz J, Reddehase MJ (1987) A nonstructural polypeptide encoded by immediate early transcription unit I of murine cytomegalovirus is recognized by cytolytic T lymphocytes. J Exp Med 166: 289–294
Landini M-P, Michelson S (1988) Human cytomegalovirus proteins. Prog Med Virol 35: 152–185
Larsen HS, Feng MF, Horohov DW, Moore RN, Rouse BT (1984) Role of T lymphocyte subsets in recovery from herpes simplex virus infection. J Virol 51: 682–686
Laubscher A, Bluestein HG, Spector SA, Zvaifler NJ (1988) Generation of human cytomegalovirus-specific cytotoxic T lymphocytes in a short-term culture. J Immunol Methods 110: 69–77
Lenardo M (1991) Interleukin-2 programs mouse $\alpha\beta$ T lymphocytes for apoptosis. Nature 353: 858–861
Ljungman P, Gleaves CA, Meyers JD (1989) Respiratory virus infection in immunocompromised patients. Bone Marrow Transplant 4: 35–40

Lukacher AE, Braciale VL, Braciale TF (1984) In vivo effector function of influenza virus-specific T lymphocyte clones is highly specific. J Exp Med 160: 814–823

Lum LG (1987) The kinetics of immune reconstitution after human bone marrow transplantation. Blood 69: 369–380

Martin PJ, Shulman HM, Schubach WH, Hansen JA, Fefer A, Miller G, Thomas ED (1984) Fatal Epstein-Barr virus-associated proliferation of donor B cells after treatment of graft versus host disease with a murine anti T cell antibody. Ann Intern Med 101: 310–315

Meyers JD, Flournoy N, Thomas ED (1980) Cytomegalovirus infection and specific cell-mediated immunity after marrow transplant. J Infect Dis 142: 816–824

Meyers JD, Flournoy N, Thomas ED (1986) Risk factors for cytomegalovirus infection after human marrow transplantation. J Infect Dis 153: 478–488

Meyers JD, Bowden RA, Counts GW (1987) Infections after bone marrow transplantation. In: Lode H, Huhn D, Melzahn M (eds) Infections in transplant patients. Thieme, Stuttgart, pp 17–32

Meyers JD, Ljungmann P, Fisher LD (1990) Cytomegalovirus excretion as a predictor of cytomegalovirus disease after marrow transplantation: importance of cytomegalovirus viremia. J Infect Dis 162: 373–380

Moore MW, Carbone FR, Bevan MJ (1988) Introduction of soluble protein into the class I pathway of antigen processing and presentation. Cell 54: 777–785

Mutter W, Reddehase MJ, Busch FW, Behring HJ, Koszinowski UH (1988) Failure in generating hemopoietic stem cells is the primary cause of death from cytomegalovorus disease in the immunocompromised host. J Exp Med 167: 1645–1658

Posavad CM, Rosenthal KL (1992) Herpes simplex virus-infected human fibroblasts are resistant to and inhibit cytotoxic T lymphocyte activity. J Virol 66: 6264–6272

Purtilo DT (1980) Epstein Barr virus induced oncogenesis in immunodeficient individuals. Lancet 1: 300–303

Quinnan GV, Kirmani N, Esber E, Saral R, Manischewitz JF, Rogers JL, Rook AH, Santos GW, Burns WH (1981) HLA-restricted cytotoxic T lymphocytes and non thymic cytotoxic lymphocyte responses to cytomegalovirus infection in bone marrow transplant recipients. J Immunol 126: 2031–2041

Quinnan GV, Kirmani N, Rook AH, Manischewitz JF, Jackson L, Moreschi G, Santos GW, Saral R, Burns WH (1982) HLA-restricted cytotoxic T lymphocytes and non thymic cytotoxic lymphocyte responses to cytomegalovirus infection. N Engl J Med 307: 6–13

Rasmussen L (1990) Immune responses to human cytomegalovirus infection. In: McDougall JK (ed) Cytomegaloviruses. Springer, Berlin Heidelberg New York, pp 221–254 (Current topics in microbiology and immunology, vol 154)

Reddehase MJ, Koszinowski UH (1984) Significance of herpse virus immediate early gene expression in cellular immunity to cytomegalovirus infection. Nature 312: 369–371

Reddehase MJ, Weiland F, Munch K, Jonjic S, Luske A, Koszinowski UH (1985) Interstitial murine CMV pneumonia after irradiation: characterization of cells that limit viral replication during established infection of the lungs. J Virol 55: 264–273

Reddehase MJ, Mutter W, Munch K, Buhring H-J, Koszinowski UH (1987a) CD8-positive T lymphocytes specific for murine cytomegalovirus immediate early antigens mediate protective immunity. J Virol 61: 3102–3108

Reddehase MJ, Mutter W, Koszinowski UH (1987b) In vivo application of recombinant interleukin-2 in the immunotherapy of established cytomegalovirus infection. J Exp Med 165: 650–656

Reddehase MJ, Rothbard JB, Koszinowski UH (1989) A pentapeptide as minimal antigenic determinant for MHC class I restricted lymphocytes. Nature 337: 651–653

Redfield RR, Wright DC, James WD, Jones TS, Brown C, Burke DJ (1987) Disseminated vaccinia in a military recruit with human immunodeficiency virus (HIV) disease. N Engl J Med 316: 673–676

Reed EC, Bowden RA, Dandliker PS, Lilleby KE, Meyers JD (1988) Treatment of cytomegalovirus pneumonia with ganciclovir and intravenous immunoglobulin in patients with bone marrow transplants. Ann Intern Med 109: 783–788

Reusser P, Riddell SR, Meyers JD, Greenberg PD (1991) Cytotoxic T lymphocyte response to cytomegalovirus following allogeneic bone marrow transplantation: pattern of recovery and correlation with cytomegalovirus infection and disease. Blood 78: 1373–1380

Riddell SR, Greenberg PD (1990) The use of anti CD3 and anti CD28 monoclonal antibodies to clone and expand antigen-specific T cells. J Immunol Methods 128: 189–197

Riddell SR, Reusser P, Greenberg PD (1991a) Cytotoxic T cells specific for cytomegalovirus: a potential therapy for immunocompromised hosts. Rev Infect Dis 13: 966–973
Riddell SR, Rabin M, Geballe AP, Britt WJ, Greenberg PD (1991b) Class I MHC-restricted cytotoxic T lymphocyte recognition of cells infected with human cytomegalovirus does not require endogenous viral gene expression. J Immunol 146: 2795–2804
Riddell SR, Watanabe KS, Goodrich JM, Li CR, Agha ME, Greenberg PD (1992a) Restoration of viral immunity in immunodeficient humans by the adoptive transfer of T cell clones. Science 257: 238–241
Riddell SR, Greenberg PD, Overell RW, Loughran TP, Gilbert MJ, Lupton SD, Agosti J, Scheeler S, Coombs RW, Corey L (1992b) Phase I study of cellular adoptive immunotherapy using genetically modified CD8$^+$ HIV-specific T cells for HIV seropositive patients undergoing allogeneic bone marrow transplant. Human Gene Ther 3: 319–338
Rosen FS, Cooper MD, Wedgewood RJP (1984a) The primary immunodeficiencies (part 1). N Engl J Med 311: 235–242
Rosen FS, Cooper MD, Wedgewood RJP (1984b) The primary immunodeficiencies (part 2). N Engl J Med 311: 300–310
Rosenberg SA, Lotze MT, Muul LM et al. (1985) Observations on the systemic administration of autologous lymphokine-activated killer cells and recombinant IL2 for patients with metastasic cancer. N Engl J Med 313: 1485–1492
Rosenberg SA, Lotze MT, Muul LM et al. (1986) A progress report on the treatment of 157 patients with advanced cancer using lymphokine-activated killer cells and interleukin-2 or high dose interleukin-2 alone. N Engl J Med 316: 1310–1321
Rosenberg SA, Packard BS, Aebersold PM, Solomon D et al. (1988) Use of tumor-infiltrating lymphocytes and IL-2 in the immunotherapy of patients with metastatic melanoma: a preliminary report. N Engl J Med 319: 1676–1680
Rosenberg SA, Aebersold P, Cornetta K, Kasid A, Morgan RA, Moen R, Karson EM, Lotze MT, Yang JC, Topalian SL, Merino MJ, Culver K, Miller AD, Blaese MR, Anderson WF (1990) Gene transfer into humans: immunotherapy of patients with advanced melanoma using tumor infiltrating lymphocytes modified by retroviral gene transduction. N Engl J Med 323: 570–578
Saral R, Burns WH, Laskin OL, Santos GW, Lietman PS (1981) Acyclovir prophylaxis of herpes simplex virus infections: a randomized, double blind controlled trial in bone marrow transplant recipients. N Engl J Med 305: 63–67
Schmidt GM, Horak DA, Niland JC, Dancan SR, Forman SJ, Zaia JA (1991) A randomized controlled trial of prophylactic ganciclovir for cytomegalovirus pulmonary infection in recipients of allogeneic bone marrow transplants. N Engl J Med 324: 1005–1011
Schrier RD, Oldstone MBA (1986). Recent clinical isolation of CMV suppress human CMV-specific human leukocyte antigen-restricted cytotoxic T lymphocyte activity. J Virol 59: 127–132
Shanley JD, Jordan MC, Cook ML, Stevens JG (1979) Pathogenesis of reactivated latent murine cytomegalovirus infection. Am J Pathol 95: 67–77
Stinski MF, Thornsen DR, Stenberg RM, Goldstein LC (1983) Organization and expression of the immediate early genes on human cytomegalovirus. J Virol 46: 1–14
Van Bleek GM, Nathenson SG (1990) Isolation of an endogenously processed immunodominant viral peptide from the class I H-2Kb molecule. Nature 348: 213–216
Van der Bruggen P, Traversari C, Chomez P, Lurquin C, De Plaen E, Van den Eynde B, Knuth A, Boon T (1991) A gene encoding an antigen recognized by cytolytic T lymphocytes on a human melanoma. Science 254: 1643–1647
Walthen M, Stinski M (1982) Temporal patterns of human cytomegalovirus transcription: mapping the viral RNAs synthesized at immediate early, early, and late times after infection. J Virol 42: 462–467
Witherspoon RP, Lum LG, Storb R (1984) Immunologic reconstitution after human marrow grafting. Semin Hematol 21: 2–10
Wold WSM, Gooding LR (1991) Region E3 of adenovirus: a cassette of genes involved in host immunosurveillance and virus-cell interactions. Virology 184: 1–8
Yewdell JW, Bennick JR (1990) The binary logic of antigen processing and presentation to T cells. Cell 62: 203–206
Yewdell JW, Bennick JR, Hosaka Y (1988) Cells process exogenous proteins for recognition by cytotoxic T lymphocytes. Science 239: 637–640
Zutter MM, Martin P, Sale GE, Shulman HM, Fisher L, Thomas ED, Durnam D (1988) Epstein Barr virus lymphoproliferation after bone marrow transplantation. Blood 72: 520–529

Cytotoxic T Lymphocytes in Human Immunodeficiency Virus Infection: Responses to Structural Proteins

R. P. JOHNSON and B. D. WALKER

1 Introduction

Following the identification of the human immunodeficiency virus type 1 (HIV-1) as the etiologic agent of acquired immunodeficiency syndrome (AIDS; BARRE-SINOUSSI et al. 1983; GALLO et al. 1984), an intensive effort has been undertaken to characterize humoral and cellular immune responses to this retrovirus. Although the immunologic correlates of protective immunity are yet to be precisely identified, accumulating evidence suggests that cytotoxic T lymphocytes (CTL) may play a central role in containing viral replication during the typically prolonged asymptomatic phase of this illness. The first reports of CTL specific for HIV appeared in 1987 (PLATA et al. 1987; WALKER et al. 1987) and documented a vigorous cellular immune response to this virus in the setting of an illness characterized by immunosuppression. Numerous studies since that time have provided detailed insight into the CTL response to HIV. Investigators have described CTL responses to multiple HIV

Infectious Disease Unit, Massachusetts General Hospital and Harvard Medical School, Boston, MA 02114, USA

Current Topics in Microbiology and Immunology, Vol. 189

proteins, identified over 40 specific epitopes, and begun to illustrate some of the functional roles of these cells.

Although HIV-specific CTL share many characteristics with CTL specific for other viruses, a number of unique features have emerged. These distinguishing features include the remarkably high frequency of activated circulating CTL, the recognition of multiple HIV epitopes, and the paradoxical persistence of viremia in the face of a vigorous virus-specific response. In this review, we will provide an overview of several aspects of HIV-1-specific CTL, including methods of detection, characterization of CTL responses to the HIV structural proteins gag, pol, and env, and a discussion of the functional role of CTL in this infection. Although our focus will be on human HIV-1-specific CTL responses, we will also discuss related findings obtained using immunized mice and rhesus monkeys infected with simian immunodeficiency virus (SIV).

2 Detection of CTL

2.1 Sources of Effector Cells

The initial demonstration of a CTL response to HIV-1 revealed a unique feature of the immune response to this infection, in that virus-specific CTL could be detected using freshly isolated lymphocytes from seropositive subjects (PLATA et al. 1987; WALKER et al. 1987). Both peripheral blood mononuclear cells (PBMC) as well as lymphocytes obtained by bronchoalveolar lavage were shown to mediate lysis of target cells bearing HIV-1 antigens, without the need for an initial period of in vitro stimulation typically required for detection of CTL responses in other viral infections studied up until that time. The vigor of the HIV-1-specific CTL response was not anticipated, particularly in light of the fact that HIV-1 infection was known to be associated with progressive and ultimately profound immunosuppression. Phenotypic studies demonstrated that lysis by these effector cells was mediated predominantly by $CD8^+$ lymphocytes and restricted by HLA class I molecules (PLATA et al. 1987; KOENIG et al. 1988; WALKER et al. 1988; KOUP et al. 1989; RIVIÈRE et al. 1989a; JOHNSON et al. 1991). An exception to this restriction pattern is the circulating envelope-specific CTL response, which is mediated by both HLA class I-restricted and nonrestricted effector mechanisms (MCCHESNEY et al. 1990; see below).The non-major histocompatibility complex (MHC)-restricted lysis of envelope-expressing targets is largely mediated by CD16-bearing lymphocytes (WEINHOLD et al. 1988; RIVIERE et al. 1989a). Although a recent report suggests the presence of CTL activity within freshly isolated CD4 lymphocytes as well as CD8 lymphocytes from infected subjects (KUNDU et al. 1992), this finding remains to be confirmed by other laboratories and appears to be in contrast with the findings of numerous other investigators.

The ability to use fresh PBMC as effector cells and thereby circumvent the usual need for in vitro, antigen-specific stimulation has greatly facilitated the characterization of human CTL responses to this virus. Not only has this characteristic permitted the identification of the major viral proteins recognized by CTL using bulk PBMC (KOENIG et al. 1988; NIXON et al. 1988; WALKER et al. 1988; KOUP et al. 1989; RIVIÈRE et al. 1989a), but in some instances the responses detected using fresh PBMC have been of sufficient magnitude to directly identify specific CTL epitopes using synthetic HIV-1 peptides to sensitize target cells (KOENIG et al. 1990b; JOHNSON et al. 1991). One can detect HIV-1-specific CTL activity using fresh PBMC in the majority of infected adults (RIVIÈRE et al. 1989b; BUSEYNE et al. 1993b). However, detection of CTL activity in unstimulated lymphocytes is relatively infrequent in children infected by vertical transmission (LUZURIAGA et al. 1991; BUSEYNE et al. 1993a).

In vitro stimulation strategies have also been utilized to detect HIV-1 specific CTL in infected subjects. Phytohemagglutinin (PHA)-stimulated lymphoblasts infected in vitro with HIV-1 have been utilized to provide antigen-specific stimulation for expansion of HIV-1-specific CTL from blood, lymph node, and lung (AUTRAN et al. 1988; HOFFENBACH et al. 1989; HADIDA et al. 1992). Stimulation of PBMC from seropositive subjects with autologous PHA-stimulated lymphoblasts without exogenous HIV infection has also resulted in boosting the in vitro measurements of HIV-1-specific CTL activity (NIXON et al. 1988, 1990). This enhancement of CTL activity has been hypothesized to occur because stimulation with PHA induces active HIV-1 replication, thereby providing an antigen-specific stimulus for expanding a memory cell population (NIXON and MCMICHAEL 1991). Others have used modifications of this approach (LIEBERMAN et al. 1992), and the resultant cell lines have provided a source of effector cells that may be used not only for functional studies, but also for pilot studies of adoptive CTL transfer in vivo (LIEBERMAN 1993; HO et al. 1993).

Specific and nonspecific cloning strategies have also been successfully employed to generate HIV-specific CTL clones from seropositive subjects, which have in turn provided the most comprehensive information regarding the spectrum of epitopes recognized of any virus studied to date. Specific stimuli have included HIV-1-infected cells (SETHI et al. 1988; HOFFENBACH et al. 1989), cells exposed to purified recombinant HIV-1 envelope protein (SETHI et al. 1988), UV-inactivated whole virus (SETHI et al. 1988), and autologous lymphocytes sensitized with peptides containing CTL epitopes (JOHNSON et al. 1992). The ability to clone HIV-1-specific CTL using polyclonal stimuli such as PHA and anti-CD3 monoclonal antibodies has greatly facilitated the precise identification of CTL epitopes recognized in natural infection and underscores the relatively high frequency of these cells in the circulating lymphocyte pool. Using monoclonal antibodies directed at the CD3 cell surface molecule, long-term cultures of virus-specific CTL have been generated in a number of studies (WALKER et al. 1989; LITTAUA et al. 1991; DAI et al. 1992; JASSOY et al. 1992; JOHNSON et al. 1991, 1992, 1993). The majority of CTL clones reported to date have been of the same phenotype and restriction pattern as the predominant CTL response observed using freshly isolated cells, namely CD8-positive, class I-restricted lymphocytes. However, at least two reports indicate the identification of CTL clones of the CD4 phenotype (LITTAUA et al. 1992; CURIEL et al. 1993), which were shown in one of these studies to be class II restricted (LITTAUA et al. 1992).

CTL responses to HIV-1 have also been identified following the in vitro stimulation of PBMC obtained from seronegative subjects (SILICIANO et al. 1988; HOFFENBACH et al. 1989) and from seronegative HIV-1 vaccine recipients (ORENTAS et al. 1990; HAMMOND et al. 1991, 1992; COONEY et al. 1993). The phenotype of cells obtained following in vitro stimulation of seronegative persons has been exclusively the CD4 phenotype and class II restricted. Both CD8-positive class I-restricted and CD4-positive class II-restricted CTL have been recovered following in vitro stimulation of PBMC from uninfected vaccine recipients (HAMMOND et al. 1992; SILICIANO et al. 1992; COONEY et al. 1993). Cellular immune responses to HIV-1 have also been studied in murine models of infection (TAKAHASHI et al. 1988) and in the SIV-infected rhesus macaque model (TSUBOTA et al. 1989; MILLER et al. 1990, 1991; SHEN et al. 1991; BOURGAULT et al. 1992; VENET et al. 1992), providing additional sources of retrovirus-specific CTL.

2.2 Sources of Target Cells

Cytotoxic T cells recognize processed viral peptides presented in association with MHC molecules (TOWNSEND et al. 1986; TOWNSEND and BODMER 1989; YEWDELL and BENNINK 1992). For $CD8^+$ CTL, these antigens consist of peptides eight to 11 amino acids in length that are presented in the cleft of a class I MHC molecule (ROTZSCHKE et al. 1990; FALK et al. 1991a; JARDETZKY et al. 1991). For $CD4^+$ CTL, antigens consist of peptides 13–17 amino acids in length that are presented by a class II MHC molecule (RUDENSKY et al. 1991; CHICZ et al. 1992). Although exceptions exist, in general, peptides presented to $CD8^+$ CTL are derived from proteins synthesized within the presenting cell (endogenous pathway), whereas peptides presented to $CD4^+$ CTL are derived from endocytosed proteins not necessarily synthesized in the presenting cell (exogenous pathway; TOWNSEND and BODMER 1989; YEWDELL and BENNINK 1992).

The requirement for the endogenous synthesis of viral antigens in target cells used to detect $CD8^+$ CTL has resulted in the use of various expression systems that express recombinant viral antigens in cells of appropriate MHC types. An additional feature that has prompted the use of recombinant expression systems is the fact that HIV-1 is a cytopathic retrovirus and therefore results in a lytic infection of CD4-bearing cells. B lymphoblastoid cell lines (B-LCL) infected with recombinant vaccinia viruses have been used in the majority of reported studies (reviewed in VENET and WALKER 1993) and have facilitated the identification of CTL responses to both structural and regulatory proteins. Tumor cell lines which have been stably transfected with plasmids expressing HIV-1 gene products have also served as a stable source of antigen-expressing target cells (PLATA et al. 1987). Autologous cell lines such as fibroblasts that have been transduced using a defective retroviral vector for the stable expression of HIV-1 genes have been used for this purpose as well (WARNER et al. 1991). Infected cells such as pulmonary alveolar macrophages (PLATA et al. 1987; HOFFENBACH et al. 1989; AUTRAN et al. 1990, 1991) and lymphocytes (HOFFENBACH et al. 1989; ORENTAS et al. 1990; TAKAHASHI et al. 1991) have served as target cells in studies that have provided evidence that HIV-1 proteins are recognized in the context of natural infection. In addition, incubation of autologous cells with purified envelope protein (SETHI et al. 1988; SILICIANO et al. 1988, 1989) or whole virus (SETHI et al. 1988) has been used to identify HIV-1-specific CTL, particularly of the CD4 phenotype.

3 CTL Specific for Human Immunodeficiency Virus Structural Proteins

3.1 Gag-Specific CTL

CTL directed at internal viral proteins such as gag are an important component of the cellular immune response to many viral infections, including HIV. Studies of murine influenza-specific CTL demonstrated that CTL were able to recognize the highly conserved internal nucleoprotein (TOWNSEND et al. 1984), and subsequent studies have shown that the nucleoprotein is the major target for cross-reactive CTL in this infection (YEWDELL et al. 1985). Although the observation that viral proteins such as the influenza nucleoprotein that are not expressed on the cell may serve as targets for immune responses was initially surprising, an improved understanding of the pathway of antigen presentation to class I-restricted CTL provided the conceptual basis for the recognition of internal viral proteins (YEWDELL and BENNINK 1992).

CTL specific for the relatively conserved HIV-1 gag protein make up a major portion of the cell-mediated responses directed against this virus. Following the initial report by WALKER et al. (1987), gag-specific CTL have been detected in PBMC of over 80% of subjects tested (NIXON et al. 1988; KOUP et al. 1989; RIVIÈRE et al. 1989a). In some subjects, remarkably high rates of lysis may be found using unstimulated PBMC as effector cells, ranging up to 80% specific lysis of gag-expressing targets using an effector-to-target ratio of 100:1 (JOHNSON et al. 1991). Cell depletion studies and analysis of inhibition by monoclonal antibodies established that the effector cells in PBMC which mediate the gag-specific CTL activity are $CD3^+$ $CD8^+$ and restricted by HLA class I molecules (NIXON et al. 1988; KOUP et al. 1989; RIVIERE et al. 1989a; JOHNSON et al. 1991). A single report suggests that $CD4^+$ CTL specific for gag may be detectable in PBMC (KUNDU and MERIGAN 1992), although this observation awaits confirmation.

These initial reports analyzed CTL activity against target cells expressing the full-length 55-kDa gag protein. However, in naturally infected cells, the 55-kDa protein is subsequently cleaved by the HIV-1 protease into smaller subunits: p24, p17, p7, and p6. Examination of CTL activity against recombinant vaccinia viruses expressing the p24 and p17 subunits has demonstrated that majority of this activity in circulating PBMC is directed against p24 (JOHNSON et al. 1991; KOUP et al. 1991b; BUSEYNE et al. 1993b). Limiting dilution analysis has also demonstrated that most of the gag-specific activity in fresh, unstimulated PBMC is directed against epitopes in p24, although increased recognition of p17 was noted in CTL precursors (KOUP et al. 1991b).

Since the initial report of a gag CTL epitope in 1988, at least 25 different peptides containing gag epitopes have been identified; they are summarized in Table 1 and Fig. 1. For a smaller number of these peptides, the minimal amino acids necessary for target cell sensitization have been defined. These epitope-mapping studies have been performed using a number of different effector cell approaches, including bulk unstimulated PBMC, PHA-stimulated PBMC, CTL clones obtained by limiting dilution, and in vitro stimulation with autologous cells superinfected with laboratory strains of virus as outlined above. The first HIV-1 CTL epitope was defined by Nixon et al., who examined the ability of PBMC cultured with autologous PHA-stimulated lymphocytes to lyse B-LCL-incubated synthetic peptides corresponding to p24 (NIXON et al. 1988). This approach has been used to identify a number of epitopes in both p24 and p17 (GOTCH et al. 1990; NIXON et al. 1988, 1990; PHILLIPS et al. 1991).

CTL epitopes in gag have also been identified using CTL clones and unstimulated PBMC. CTL clones developed by the limiting dilution culture of lymphocytes stimulated with a CD3-specific monoclonal antibody, recombinant IL-2, and irradiated allogeneic PBMC from seronegative donors (WALKER et al. 1989) have proved very useful for the identification and detailed characterization of CTL epitopes. These CTL clones can be maintained in long-term tissue culture and frozen and thawed without loss of CTL activity and have made possible the fine mapping of the specificity of CTL clones and analysis of the effects of sequence variation (JOHNSON et al. 1991, 1993). A total of eight CTL epitopes in p17 and p24 were defined in this fashion. A particularly notable feature was the recognition of multiple gag epitopes (up to four) by the same individual and the presentation of more than one gag epitope by the same HLA molecule. Reports by other investigators have described additional epitopes using this technique as well. In selected patients, vigorous gag-specific responses in unstimulated PBMC have also allowed the epitopes to be identified without in vitro stimulation or culture (JOHNSON et al. 1991). Although the vast majority of the gag-specific CTL clones have been $CD8^+$ and restricted by HLA class I antigens, a single report exists of a $CD4^+$, HLA class II-restricted CTL clone which recognizes an epitope in p24 (LITTAUA et al. 1992). The relative rarity of such clones and the fact that most investigators have found that bulk PBMC responses have been shown to be restricted by class I antigens (NIXON et al. 1988; KOUP et al. 1989; RIVIERE et al. 1989a; JOHNSON et al. 1991) suggest that the overall contribution of $CD4^+$ CTL to the detectable gag-specific response is relatively small.

Alternative approaches to the identification of CTL epitopes include the use of peptide binding assays and of CTL prediction algorithms. Peptide binding assays have been derived in which synthetic HIV peptides have been tested for their ability to bind to purified HLA molecules attached to a solid support. This approach has been used by several groups to identify viral peptides that bind to MHC molecules (CHOPPIN et al. 1990, 1991a, b; FRELINGER et al. 1990). However, only a subset of peptides identified in this fashion are known to contain CTL epitopes. The application of an algorithm which predicts the existence of CTL epitopes based on the relative rarity of tetrapeptide sequences in comparison to sequenced human proteins has been described by CLAVERIE et al. (1988). Four epitopes were identified in this fashion, two lying in p24 and one each in p7 and p6. Lines from HIV-1-seropositive subjects were prepared by in vitro stimulation of PBMC with autologous HIV-infected cells and tested for lysis of autologous cells incubated with peptides predicted by the algorithm. The four epitopes predicted in this fashion were shown to sensitize targets for lysis by effector cells prepared using this technique, and lysis appeared to be restricted by HLA-A2. However, none of these epitopes has yet been confirmed by other investigators and most of the peptides do not clearly exhibit the characteristics of a recently proposed HLA-A2 binding motif (FALK et al. 1991b).

The gag-specific CTL response has also been investigated in the SIV macaque model of infection (VOWELS et al. 1989; MILLER et al. 1990; YAMAMOTO et al. 1990b; VENET et al. 1992). Detection of SIV-specific CTL has generally required in vitro expansion in the presence of a stimulus such as Con A and has not been detectable in freshly isolated PBMC. Cytolytic activity has been identified in lymphocytes derived from PBMC as well as lymph nodes (REIMANN et al. 1991) and skin lesions (YAMAMOTO et al. 1992). A dominant nine-amino acid gag epitope (aa 182–190 of SIV mac251) recognized by MHC class I-restricted CTL has been characterized, and the restricting antigen has been sequenced and termed Mamu-A*01

Table 1. Human immunodeficiency virus type 1 gag peptides recognized by human cytotoxic T lymphocytes

Figure number	HLA	Amino acids	Sequence	Reference
1	A2	71–85	GSEELRSLYNTVATL	Nixon and McMichael 1991
1	A2	77–85	SLYNTVATL	Johnson et al. 1991 and unpublished data
2	A2	88–115	VHQRIEIKDTKEALDKIEEEQNKSKKKA	Achour et al. 1990
3	A2	193–203	GHQAAMQMLKE	Claverie et al. 1988
4	A2	219–233	HAGPIAPGQMREPRG	Claverie et al. 1988
5	A2	418–433	KEGHQMKDCTERQANF	Claverie et al. 1988
6	A2	446–460	GNFLQSRPEPTAPPF	Claverie et al. 1988
7	A3	18–31	KIRLRPGGKKKYKL	Jassoy et al. 1992
8	A11	84–92	TLYCVHQRI	T. Harrer, E. Harrer, B.D.W., unpublished data
9	A11	343–363	LEEMMTACQGVGGPGHKARVL	R.P.J. and B.D.W, unpublished data
10	A33	263–277	KRWIILGLNKIVRMY	Buseyne et al. 1993b
11	B8	25–35	GKKKYKLHIV	Phillips et al. 1991
12	B8	253–267	NPPIPVGEIYKRWII	Gotch et al. 1990
12	B8	253–274	NPPIPVGEIYKRWIILGLNKIV	Johnson et al. 1991
12	B8	256–270	IPVGEIYKRWIILGL	Buseyne et al. 1993b
13	B8	313–334	VKNWMTETLLVQNANPDCKTIL	Johnson et al. 1991
13	B8	323–339	VQNANPDCKYILKAL	Nixon and McMichael 1991
14	B12(44)	169–184	IPMFSALSEGATPQDL	Buseyne et al. 1993b
15	B14	173–194	SALSEGATPQDLNTMLNTVGGH	Johnson et al. 1991
15	B14	181–191	PQDLNTMLNTV	Koenig et al. 1990a
15	B14	183–189	DLNTMLNTVGGHQAA	Nixon and McMichael 1991
16	B14	305–313	RAEQASQEV	Johnson et al. 1991
17	B27	263–277	KRWIILGLNKIVRMY	Nixon et al. 1988
17	B27	263–272	KRWIILGLNK	Buseyne et al. 1993b
18	Bw52	193–214	GHQAAMQMLKETINEEAAEWDR	Johnson et al. 1991
19	Bw57	143–164	VHQAISPRTLNAWVKVVEEKAF	Johnson et al. 1991
20	Bw57	153–174	NAWVKVVEEKAFSPEVIPMFSA	Johnson et al. 1991
21	Bw62	18–31	KIRLRPGGKKKYKL	Johnson et al. 1991 and unpublished data
22	Bw62	268–278	LGLNKIVRMYS	Johnson et al. 1991 and unpublished data
23	Cw3	145–150	QAISPR	Littaua et al. 1991
24	DQw1	140–148	GQMVHQAIS	Littaua et al. 1992
25	DQw3	140–148	GQMVHQAIS	Littaua et al. 1992

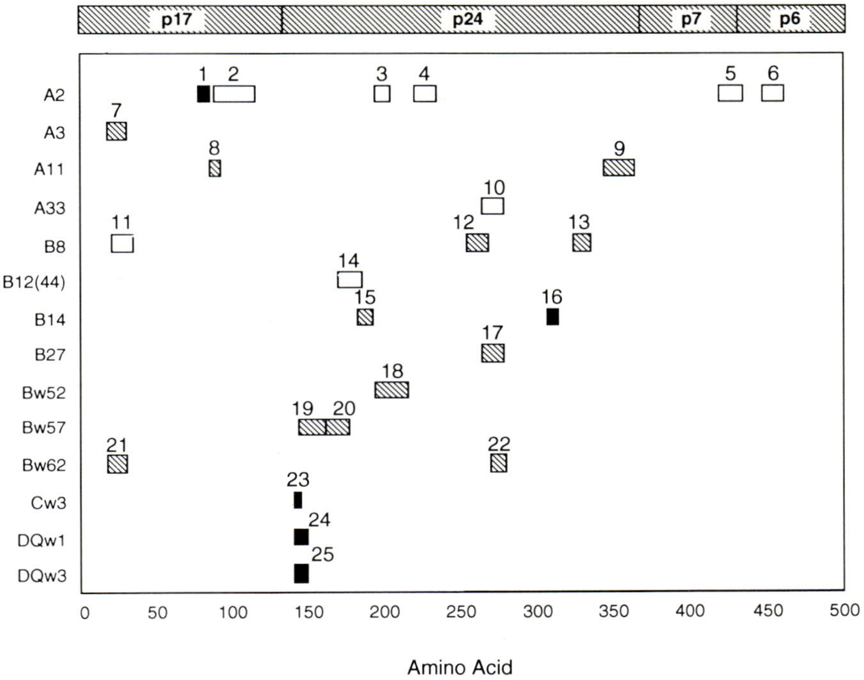

Fig. 1. Human immunodeficiency virus type 1 (HIV-1) gag peptides recognized by human cytotoxic T lymphocytes (CTL). Only epitopes with defined major histocompatibility complex restriction have been included. *Open bars* indicate epitopes that have been defined using polyclonal effector cells and not yet confirmed by other investigators. *Hatched bars* indicate epitopes that have been defined using polyclonal effectors and confirmed by other investigators or epitopes defined using CTL clones. *Black bars* indicate epitopes that have been finely mapped to minimal epitopes. *Numbers* refer to epitope sequences listed in Table 1. Numbering of amino acids corresponds to the HXB2R molecular clone (MYERS et al. 1991)

(YAMAMOTO et al. 1990a). Because all Mamu-A*01-positive animals appear to generate a CTL response to this epitope, a variety of immunization strategies are presently being evaluated in Mamu-A*01 animals.

Although the identification of HIV-1 gag CTL epitopes is likely to be incomplete, several general themes have emerged. The gag-specific response in HIV-1 infected humans appears to be remarkably heterogeneous. Multiple epitopes have been identified, and as many as five gag CTL epitopes have been shown to be recognized by a single individual (JOHNSON et al. 1991 and unpublished data). Multiple HLA types may be utilized to present gag epitopes, even in the same individual, and the same HLA type may in turn present more than one gag epitope. Similar findings have also been noted for other HIV-1 proteins such as reverse transcriptase (RT) and nef (see below and accompanying article by Rivière). An example of the breadth and vigor of

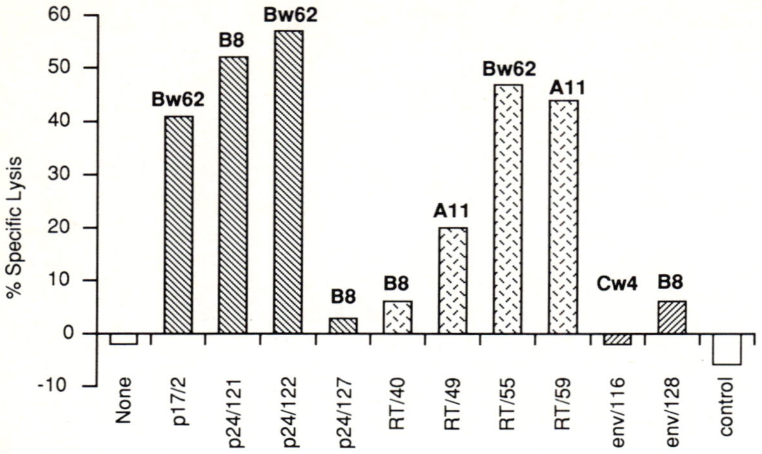

Fig. 2. Recognition of multiple human immunodeficiency virus type 1 (HIV-1) cytotoxic T lymphocyte (CTL) epitopes by unstimulated peripheral blood mononuclear cells (PBMC) from a HIV-1-seropositive donor. Fresh, unstimulated PBMC were used as effector cells at an effector-to-target ratio of 100:1 in a 6-h ^{51}Cr-release assay. Target cells consisted of autologous B lymphoblastoid cell line (B-LCL) that had been sensitized with peptides (100 μg/ml) previously demonstrated to contain CTL epitopes (WALKER et al. 1988; JOHNSON et al. 1991, 1992, 1993). The HLA restriction element for each peptide is indicated above the *bar*

the HIV-1-specific CTL response is shown in Fig. 2, in which unstimulated PBMC from an HIV-1-seropositive subject were shown to recognize at least six CTL epitopes. A total of 14 distinct HIV-1 CTL epitopes have been identified in this subject (WALKER et al. 1988; JOHNSON et al. 1991, 1992 and unpublished data). These results stand in marked contrast to the observations obtained regarding virus-specific CTL responses in inbred mouse strains. In these systems, CTL responses to viruses are dominated by one or two immunodominant epitopes (BRACIALE et al. 1989), and the absence of the appropriate MHC type may result in a failure to mount a CTL response to a particular virus (BENNINK and YEWDELL 1988). It remains to be seen whether these differences are related to the genetic heterogeneity of humans as opposed to the inbred mice or whether this is related to the chronic antigenic stimulation observed with HIV-1 infection. Another intriguing observation is the presence of overlapping CTL epitopes. Although epitopes are distributed throughout gag, several "hot spots" exist, where a series of overlapping CTL epitopes have been observed, each presented by a different HLA type. These regions can be identified in Fig. 1. In several cases, the epitopes have been finely mapped and shown to consist of distinct overlapping epitopes, as has been observed for nef (CULMANN et al. 1989, 1991) and gp41 (JOHNSON et al. 1992). The basis for such overlapping epitopes remains unclear. Possibilities include common pathways of antigen processing of peptides or chance association.

3.2 Reverse Transcriptase-Specific CTL

The HIV-1 polymerase gene encodes the viral RT protein, which is essential for replicative function. Because this protein exhibits the least variation among sequenced isolates, CTL responses to this protein might be expected to be highly cross-reactive with a number of different isolates and therefore important in vaccination strategies designed to elicit cross-protective immunity. As has previously been shown for the influenza virus polymerase (BENNINK et al. 1982), the HIV-1 polymerase is a major target of virus-specific CTL (WALKER et al. 1988). Freshly isolated PBMC from infected persons in various stages of HIV-1 infection were found to lyse autologous B-LCL infected with recombinant vaccinia viruses expressing the RT protein, and this lysis is restricted by class I antigens (WALKER et al. 1988).

Specific epitopes recognized by RT-specific CTL have been identified using cloned CTL from infected subjects, and these studies demonstrate that the CTL response to this protein is directed at a number of different epitopes (Table 2, Fig. 3). In one study, PBMC from two subjects with high levels of RT-specific lysis were stimulated in vitro with a CD3-specific monoclonal antibody in the presence of irradiated allogeneic feeder cells and interleukin 2 and cloned at limiting dilution (WALKER et al. 1989). Using recombinant HIV-1 vaccinia viruses expressing truncated portions of the RT protein, regions containing CTL epitopes were identified, and fine mapping of these epitopes was achieved using synthetic HIV-1 RT peptides. Five different peptides containing RT CTL epitopes were identified throughout the protein and found to be restricted by multiple different HLA class I antigens. In one of these subjects, two different epitopes restricted by the same class I antigen were identified. No epitopes were recognized by both individuals. Examination of the peptides containing these CTL epitopes revealed a high degree of homology among sequenced isolates, suggesting that CTL responses to the RT might be cross-reactive with multiple strains. This initial study indicated that the CTL response to a conserved retroviral protein is highly heterogeneous in infected persons and suggested that the immunogenic viral epitopes are different in different individuals. Similar heterogeneity was observed in another report on RT-specific CTL, in which multiple regions of the RT protein were found to contain CTL epitopes recognized using in vitro expanded lymphocytes from four seropositive subjects (LIEBERMAN et al. 1992).

Other studies investigating the CTL response to HIV-1 RT have utilized a murine system in which mice were immunized with a recombinant vaccinia-RT virus (HOSMALIN et al. 1990). In contrast to the heterogeneous response to HIV-1 RT which had been observed in humans (WALKER et al. 1988; LIEBERMAN et al. 1992), a single immunodominant epitope distinct from previously reported epitopes was identified in the immunized mice, located in the N-terminal region of the RT. When autologous target cells sensitized with the peptide containing this epitope were tested in a small cohort of infected

Table 2. Human immunodeficiency virus type 1 pol cytotoxic T lymphocyte epitopes recognized by human CTL

Figure number	HLA	Amino acids	Sequence	Reference
1	A2	464–472	I L K E P V H G V	Walker et al. 1988; Tsomides et al. 1991
2	A11	313–321	A I F Q S S M T K	Walker et al. 1988 and unpublished data
3	A11	496–507	Q I Y Q E P F K N L K T G	Walker et al. 1988 and unpublished data
4	A28	585–609	E K E P I V G A E T F Y V D G A A N R E T R L G K	R.P.J., B.D.W., unpublished data
5	B8	160–184	I E T V P V K L K P G M D G P K V K Q W P L T E E	Walker et al. 1988
6	B14	636–660	A I Y L A L Q D S G L E V N I V T D S Q Y A L G I	R.P.J., B.D.W., unpublished data
7	Bw60	355–366	T K I E E L R Q H L L R	Walker et al. 1988 and unpublished data
8	Bw62	415–426	L V G K L N W A S Q I Y	Walker et al. 1988 and unpublished data
9	Bw62	464–473	I L K E P V H G V Y	R.P.J., B.D.W., unpublished data

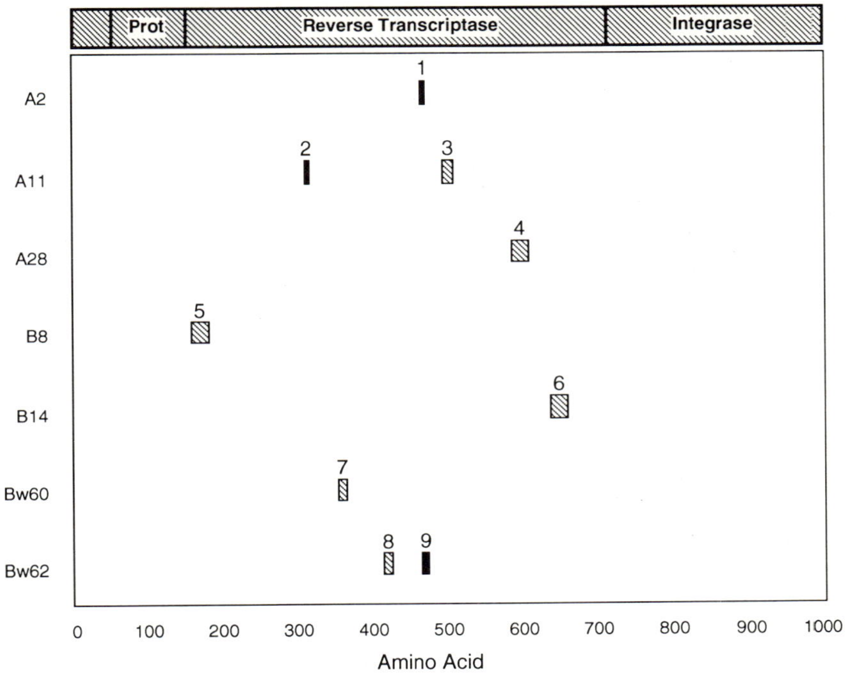

Fig. 3. Human immunodeficiency virus type 1 (HIV-1) reverse transcriptase peptides recognized by human cytotoxic T lymphocytes (CTL). *Prot,* protease. *Numbers* refer to epitope sequences listed in Table 2. See also legend to Fig. 1

subjects using freshly isolated lymphocytes as effector cells, consistent lysis was observed in three of 12 patients. Although lysis could be inhibited by anti-CD3 and anti-CD8 monoclonal antibodies, there did not appear to be any correlation between any HLA antigen and responsiveness to this epitope.

3.3 Envelope-Specific CTL

The HIV-1 envelope protein is also a target for CTL in infected subjects, although characterization of this response has proved complicated for a number of reasons. Analysis of CTL activity directed against the HIV-1 envelop glycoprotein gp160 has revealed the presence of both HLA-restricted and MHC-unrestricted lysis of env-expressing targets in PBMC. The initial report by WALKER et al. (1987), demonstrating inhibition of envelope-specific lysis with an anti-CD3 monoclonal antibody, suggested that the effector cells in PBMC were T lymphocytes, although data on restriction of lysis by HLA antigens were not definitive. Subsequent reports from other groups also demonstrated difficulties in precisely characterizing the effector cells of the envelope-specific response. KOENIG et al. (1988) noted that lysis of env-expressing target cells by PBMC was not MHC restricted, although it could be completely inhibited by anti-CD3 monoclonal antibody and partially inhibited by anti-CD8 monoclonal antibody. Similarly, RIVIÈRE and colleagues (1989a) were unable to demonstrate MHC-restricted lysis of env-expressing targets using PBMC and suggested that envelope-specific, antibody-dependent cellular cytotoxicity (ADCC) might be a second effector mechanism. Using PBMC depleted of NK cells, KOUP et al. (1989) demonstrated lysis of env-expressing targets in an HLA class I-restricted fashion, consistent with the identity of effector cells as CTL. Additional early evidence supporting the existence of MHC-restricted CTL mediating lysis of env-expressing targets was also provided by KOENIG et al. (1988), who derived CTL clones by limiting dilution, which were shown to be $CD8^+$ and HLA class I restricted.

These reports suggested the possibility that killing against envelope-expressing targets was being mediated by two different effector cell populations, one HLA-restricted and the other non-HLA-restricted. In order to distinguish killing of env-expressing target cells by the two populations of effector cells, MCCHESNEY et al. (1990) utilized recombinant vaccinia viruses which expressed gp160 in either an unmodified form or in a form which had a deletion in the signal peptide. The mutant envelope construct lacking the signal peptide was not expressed on the cell surface and could therefore not serve as a target for ADCC. Using these constructs to assess CTL activity and total envelope-specific cytolytic activity, the authors noted that fresh PBMC from only about one third of subjects tested were able to lyse autologous targets infected with the signal deletion construct, although 92% of subjects were able to recognize the unmodified gp160-expressing targets. Confirmatory data have been obtained in studies performed on rhesus monkeys infected with SIV (YAMAMOTO et al. 1990a) that support the existence of both ADCC mediating MHC-unrestricted killing and CTL mediating MHC-restricted killing. A detailed characterization of the HLA-unrestricted lysis of envelope-expressing targets has been provided by TYLER et al. (1989), who described a $CD3^-$ and $CD16^+$ cell population which was able to lyse target cells expressing gp160 in an MHC-unrestricted fashion. Cytophilic antibody specific for gp 120 could be dissociated from these cells. These data suggest that the HLA-unrestricted lysis against envelope-expressing cells consist predominantly of natural killer and killer cells which are armed in vivo with antibodies against gp160. Thus, the cytolytic response in PBMC appears to consists primarily of two populations of cells: classical CTL, which are $CD3^+$ $CD8^+$ and restricted by HLA class I antigens, and armed killer cells, which are $CD3^-$ $CD16^+$.

The existence of both HLA-restricted and -unrestricted killing against the HIV envelope has complicated the identification of CTL epitopes in this protein. Owing in part to this difficulty, the initial descriptions of envelope CTL epitopes occurred using experimental systems not employing CTL from

HIV-1-seropositive subjects. TAKAHASHI et al. (1988) analyzed CTL derived from BALB/c mice that had been immunized with recombinant vaccinia virus containing gp160 derived from the IIIB strain. CTL derived from these mice following antigenic stimulation in vitro were $CD8^+$ and MHC restricted. Epitope mapping demonstrated that the dominant response was directed against a single epitope in a highly variable portion of gp160, the V3 loop, which also contains one of the major epitopes recognized by neutralizing antibodies. This initial response demonstrating an immunodominant CTL response in mice directed against a highly variable portion of the envelope raised the disturbing possibility that HIV-1 envelope-specific CTL would be largely type specific. Using a different technique, SILICIANO et al. (1988) derived $CD4^+$ envelope-specific CTL from PBMC from HIV-1-seronegative donors using in vitro stimulation with gp120 and a soft agar gel cloning technique. The specificity of one of these clones was mapped to a relatively conserved region of gp120. However, a single conservative amino acid substitution was able to abrogate recognition of the variant peptide (SILICIANO et al. 1988; CALLAHAN et al. 1990).

In recent years, a number of CTL epitopes have been derived using CTL clones and PBMC from HIV-1-infected humans (Table 3, Fig. 4). Epitopes identified using $CD8^+$ CTL clones have been largely in gp41 (TAKAHASHI et al. 1991; DAI et al. 1992; JOHNSON et al. 1992), although epitopes also have been identified in gp120 (JOHNSON et al. 1993 and R.P.J., S. Hammond, R. Siliciano and B.D.W., manuscript in preparation). A particularly interesting region lies in the extracellular domain of gp41 immediately adjacent to the immunodominant B cell epitope. This region contains a ten-amino acid stretch which contains CTL epitopes recognized by HLA-A28-, B8-, and B14-restricted CTL (DAI et al. 1992; JOHNSON et al. 1992) and overlaps with a CTL epitope recognized by $CD4^+$ CTL obtained from HIV-1-seronegative volunteers who had been immunized with rgp160 (HAMMOND et al. 1991). Reports using unstimulated PBMC or PBMC stimulated with HIV-infected lymphoblasts have suggested the existence of multiple CTL epitopes distributed throughout gp120 and gp41, lying both in conserved and variable regions of gp160 (CLERICI et al. 1991; DADAGLIO et al. 1991).

These results from the studies described above suggest that there are a number of potential CTL epitopes in gp160 and that these epitopes lie in both conserved and variable regions. However, interpretation of the data obtained from HIV-1-seropositive subjects is complicated by the fact that detection of CTL activity is assessed using target cells expressing HIV-1 envelope derived from laboratory strains of HIV-1. Because of the sequence variation observed in HIV-1 envelope, which may approach 30% at the amino acid level, CTL assays conducted using HIV-1 proteins derived from laboratory strains are likely to underestimate the total CTL response. Recent data from our laboratory using CTL obtained from a laboratory worker who was accidentally infected with HIV-1 IIIB help to address this issue (R. P. Johnson, W. A.

Table 3. Human immunodeficiency virus (HIV) type 1 envelope peptides recognized by human cytotoxic T lymphocytes

Figure number	HLA	Amino acids	Sequence	Reference
1	A2	034–055	L W V T V Y Y G V P V W K E A T T T L F C A	Dadaglio et al. 1991
2	A2	112–124	W D Q S L K P C V K L T P	Clerici et al. 1991
3	A2	188–207	T T S Y T L T S C N T S V I T Q A C P K	Dadaglio et al. 1991
4	A2	291–307	S V E I N C T R P N N N T R K S I	Dadaglio et al. 1991
5	A2	308–322	R I Q R G P G R A F V T I G K	Clerici et al. 1991; Dadaglio et al. 1991
6	A2	369–375	P E I V T H S	Dadaglio et al. 1991
7	A2	377–387	(K) N G G E F F Y C N S	Dadaglio et al. 1991
8	A2	416–435	L P C R I K Q F I N M W Q E V G K A M Y	Dadaglio et al. 1991
8	A2	421–436	K Q I I N M W Q E V G K A M Y A	Clerici et al. 1991
9	A2	489–508	V K I E P L G V A P T K A K R R V V Q R	Dadaglio et al. 1991
10	A2	827–841	D R V I E V V Q G A Y R A I R	Clerici et al. 1991
11	A3	37–46	T V Y Y G V P V W K	R.P.J, S. Hammond, R. Siliciano, B.D.W., manuscript in preparation
12	A3	770–780	R L R D L L L I V T R	Takahashi et al. 1991
13	A24	770–780	R L R D L L L I V T R	R. Koup, personal communication
14	A24	586–593	Y L K D Q Q L L	Dai et al. 1992
15	A30	837–856	Y R A I R H I P R R I R Q G L E R I L L	Lieberman et al. 1992
16	B8	586–593	Y L K D Q Q L L	Johnson et al. 1992
17	B8	837–856	Y R A I R H I P R R I R Q G L E R I L L	Lieberman et al. 1992
18	B14	584–592	E R Y L K D Q Q L	Johnson et al. 1992
19	B27	781–802	I V E L L G R R G W E A L K Y W W N L L Q Y	Lieberman et al. 1992
20	B35	592–616	L L G I W G C S G K L I C T T A V P W N A S W S N	R.P.J, S. Hammond, R. Siliciano, B.D.W., manuscript in preparation
21	Cw4	376–383	F N C G G E F F	Johnson et al. 1993
22	DR4	410–429	G S D T I T L P C R I K Q F I N M W Q E	Siliciano et al. 1988
23	DPw4.2	579–590	R I L A V E R Y L K D Q	Hammond et al. 1991

Pepide amino acids enclosed in parentheses do not correspond to sequenced HIV isolates.

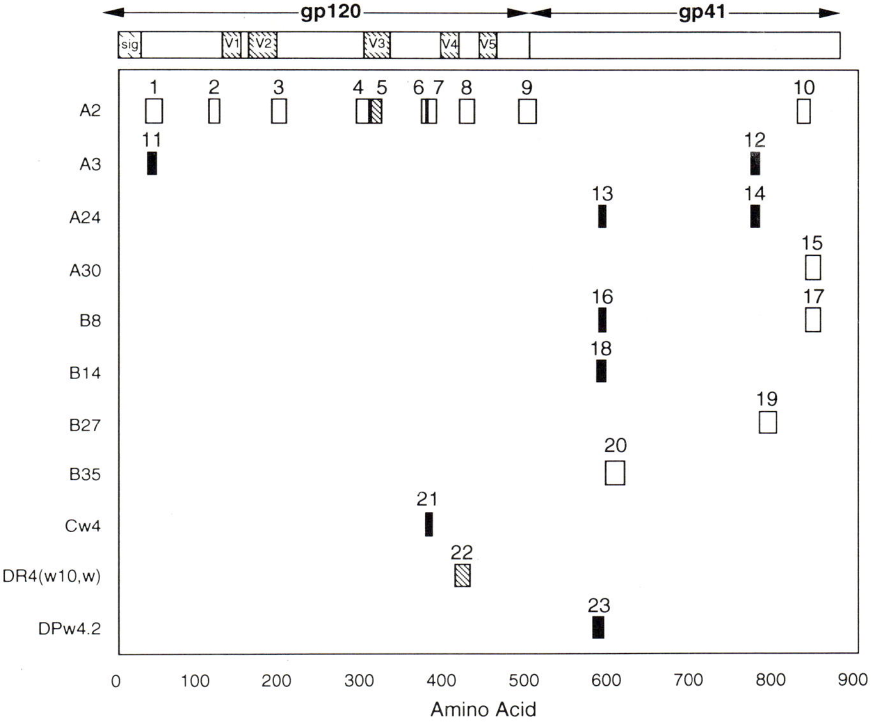

Fig. 4. Human immunodeficiency virus type 1 (HIV-1) envelope peptides recognized by human cytotoxic T lymphocytes (CTL). *Sig*, signal peptide; *gp*, glycoprotein. The five hypervariable regions of gp120 are indicated *V1–V5*. *Numbers* refer to epitope sequences listed in Table 3. See also legend to Fig. 1

Blattner, manuscript in preparation). CTL clones derived from this subject and screened for CTL activity against target cells expressing HIV-1 IIIB proteins suggest that the CTL response is directed against both relatively conserved and variable regions of gp160. In the only subject studied so far, CTL activity would have been underestimated by approximately half, had target cells not expressed autologous viral proteins.

4 Functional Aspects of CTL

Although HIV-1 infection is associated with a vigorous CTL response, this response is clearly insufficient to eliminate the virus from the infected host. A number of functional studies have now begun to more specifically address the functional role of CTL in HIV-1 infection.

4.1 Production of Cytokines by CTL

HIV-1-specific CTL not only lyse infected cells, but are also induced to release a number of inflammatory cytokines upon recognition of their specific target epitopes. Both cerebrospinal fluid (CSF)-derived as well as PBMC-derived CTL clones have been shown to release γ-interferon (IFN-γ), tumor necrosis factor alpha (TNF-α), and TNF-β in an epitope-specific and HLA class I-restricted fashion, paralleling the HIV-1-specific cytolytic activity of these cells (JASSOY et al. 1993). These studies suggest that the elevated levels of these inflammatory cytokines that are found in the serum and CSF of patients with HIV-1 infection may be due at least in part to the vigorous HIV-1-specific CTL response and may have important implications for understanding HIV-1 disease pathogenesis, providing a physiologic link between the virus-specific immune response and some of the neuropathologic sequelae of HIV-1 infection. For example, TNF-α may contribute as a local host defense by directly inhibiting HIV-1 replication (HARTSHORN et al. 1987) or by upregulating class I expression, leading to augmented recognition of infected target cells (VIDOVIC et al. 1990; JOLY et al. 1991; SETHNA and LAMPSON 1991). Alternatively, IFN-γ, TNF-α, and TNF-β have been shown to be capable of inducing HIV-1 expression in promonocyte and T cell lines in vitro (MATSUYAMA et al. 1991; BISWAS et al. 1992), and TNF-α release by HIV-1-specific CTL has been shown to be sufficient to upregulate HIV-1 replication in chronically infected bystander cells (HARRER et al. 1993). In addition, release of cytokines by CTL may contribute to the pathogenesis of HIV-1-related neurologic disorders such as AIDS dementia complex (ADC). A vigorous CTL response has been detected in the CSF of persons with ADC, and these cells release IFN-γ, TNF-α, and TNF-β upon recognition of their respective target cells (JASSOY et al. 1992). IFN-γ has been shown to induce metabolic changes in macrophages and microglial cells, leading to altered metabolic pathways for kynurinine and tryptophan, which may lead to elevated levels of neopterin and quinolonic acid, which have been associated with HIV-1 infection (WERNER et al. 1987; HEYES et al. 1991). TNF-α and TNF-β stimulate astrocytes to proliferate and may result in reactive gliosis (SELMAJ and RAINE 1988), and TNF-α has also been linked to demyelination in the central nervous system (CNS; SELMAJ et al. 1990). These histopathologic changes are typically seen in HIV-1-related neurologic disorders and suggest that the state of immune activation may be related to ultimately deleterious effects.

4.2 CTL and Immune Escape

The paradox of ongoing HIV replication in the face of a vigorous virus-specific CTL has prompted investigation into potential mechanisms used by HIV to avoid suppression by, and to eventually escape from, the host immune response. One leading hypothesis to explain the escape of HIV from immune control postulates that sequence variation in HIV results in the evolution of viruses that are not recognized by the host immune response. HIV isolates exhibit extensive variation, as do other lentiviruses. Variation occurs even in a single subject, leading to the concept that HIV-1-seropositive patients are infected with a mixture of related viruses, referred to as quasispecies. According to this hypothesis, viruses that have developed mutations in CTL epitopes that decrease recognition by CTL would fail to be recognized and have a selective advantage in vivo. Some evidence to support sequence variation as a mechanism for escape already exists with regard to humoral immunity (REITZ et al. 1988; ALBERT et al. 1990; NARA et al. 1990). However, available data have not yet definitively established a link between sequence variation and escape from antibodies directed against HIV-1.

Several reports have addressed the relationship between sequence variation in vivo and recognition by CTL. PHILLIPS et al. (1991) examined CTL responses and sequence variation in a cohort of HIV-1-seropositive hemophiliacs. HIV-1 gag CTL epitopes were identified which were presented by either HLA-B8 or HLA-B27, and CTL responses to these epitopes were evaluated longitudinally. In addition, segments of HIV-1 DNA coding for these CTL epitopes were amplified using the polymerase chain reaction (PCR), sequenced, and variant peptides corresponding to the deduced amino acid sequences tested for recognition by CTL. Examination of sequence variation occurring in HLA-B8-restricted CTL epitopes in p17 and p24 revealed the appearance of mutations which resulted in either a loss of recognition by CTL or, in one instance, a shift in recognition of CTL from one peptide to a variant peptide. The relative frequency of viruses containing the variant sequence varied over time, but there was no clear selection over time of viruses containing mutant sequences which were not recognized. Sequence variation was also observed in an HLA-B27-restricted CTL epitope in p24, but in contrast to the HLA-B8-restricted CTL epitopes, this variant did not result in a loss of sequence variation. Using a similar approach, MEYERHANS et al. (1991) observed the appearance of sequence variation in the same HLA-B27-restricted CTL epitope, but did not observe that any of these changes resulted in a loss of recognition by CTL. The relationship of sequence variation and recognition by CTL has also been addressed in the SIV model (CHEN et al. 1992). As noted above, the SIV-gag-specific CTL response in infected rhesus macaques expressing the Mamu-A*01 MHC type is directed against a single immunodominant epitope (amino acids 182–190 in the SIV p28 gap protein; MILLER et al. 1991). CHEN et al. (1992) examined sequence variation in this epitope in three SIV-infected monkeys expressing the Mamu-A*01 haplotype. Sequence variation occurred within the CTL epitope, but did not appear to occur more frequently within the CTL epitope than in surrounding areas. Furthermore, there did not appear to be a significantly increased frequency of nonsynonymous nucleotide substitutions (i.e., nucleotide changes which result in amino acid substitutions) compared with synonymous substitutions. None of the observed amino acid substitutions resulted in a loss of recognition by CTL, although the CTL utilized in these assays of variant peptides were not derived from the monkeys from whom the sequence data were derived.

These conflicting data do not unequivocally support the hypothesis that sequence variation is a major mechanism of escape from CTL. The data obtained by PHILLIPS et al. (1991) as well as unpublished data obtained from our laboratory suggest that variant viruses may appear which are not recognized by CTL and that in some cases, a shift in CTL specificity may occur. However, this does not always appear to be the case, as evidenced by the data from other groups (MEYERHANS et al. 1991; CHEN et al. 1992). It is possible that the failure to observe sequence variants arising which escape CTL recognition for certain CTL epitopes may reflect structural constraints on the virus that limit its ability to mutate in a specified region. Critical evidence that is lacking thus far is a clear demonstration that viruses with mutations in CTL epitopes have a selective advantage in vivo.

Mechanisms other than sequence variation which may allow HIV to escape CTL responses have been proposed, but specific data in favor of these hypotheses is lacking at this time. HIV-1 infection of lymphocytes has been observed to produce a transient decrease in the expression of MHC class I molecules in transformed cell lines and PHA-stimulated lymphocytes (KERKAU et al. 1989; SCHEPPLER et al. 1989). This downregulation of HLA class I expression was shown to decrease killing by allogeneic CTL, but no specific data have been published regarding the effects on killing by HIV-1-specific CTL. Decreased expression of lymphocyte function associated antigen-1 (LFA-1) on PBMC obtained from HIV-1 seropositive subjects has also been reported (AMIEL et al. 1988). However, the ability of HIV-1-specific $CD8^+$ CTL to kill HIV-infected lymphocytes (HOFFENBACH et al. 1989; ORENTAS et al. 1990; TAKAHASHI et al. 1991 and RPJ, BDW, unpublished data) suggests that this downregulation of selected cell surface markers does not prevent recognition of infected cell lines by HIV-1-specific CTL, and further experiments are necessary to define the role that downregulation of cell surface molecules on infected lymphocytes may play. Other potential mechanisms of immune escape from CTL that have been observed in other viral infections, such as selective blocks in antigen presentation to the class I pathway (DEL VAL et al. 1989), may also be operative in HIV-infected cells, but there is no specific evidence for this possibility at the present time.

4.3 Inhibition of Human Immunodeficiency Virus Type 1 Replication by CTL

The ability of CTL to lyse cells expressing viral antigens in in vitro assays suggests that these cells are likely to be active in vivo in inhibiting viral replication. Evidence for the protective role of CTL against viral infection in general arises from the observations that adoptive transfer of virus-specific CTL, including influenza, lymphocytic choriomeningitis virus, and murine cytomegalovirus, is effective in preventing viral infection in naive mice. However, only preliminary evidence has appeared to date which addresses the ability of HIV-1-specific CTL to inhibit viral replication. Preliminary data from other investigators (KOENIG et al. 1991) and unpublished data from our group (R.P.J., V. Johnson and B.D.W.) have demonstrated that HIV-1-specific CTL clones are able to inhibit viral replication in vitro. This inhibition of viral replication is largely HLA restricted, but there is also a component which is not HLA restricted. The non-HLA-restricted inhibition of viral replication is mediated, in part, by the release of soluble factors which may be related to the soluble inhibitory factors produced by $CD8^+$ lymphocytes identified in previous studies by C. WALKER and colleagues (1986; WALKER and LEVY 1989). The ability of these CTL clones to inhibit viral replication has also been studied in adoptive transfer experiments carried out in the severe combined immunodeficiency (SCID)-hu-PBMC model. In this system, mice with SCID are reconstituted with PBMC and then subsequently infected with HIV-1. Preliminary results utilizing the transfer of nef-specific CTL clones (KOENIG et al. 1991, 1992) and RT-specific CTL clones (KOUP et al. 1991a) into HIV-1-infected SCID-hu-PBMC mice suggests that these clones are also effective in inhibiting viral replication in this animal model, although both HLA-restricted and non-HLA-restricted inhibition of viral replication has been observed. Clearly, further research is needed in order to more precisely characterize the relative contributions of cytolytic and noncytolytic pathways in inhibiting viral replication.

4.4 Vaccine-Related Issues

At least 14 different HIV-1 vaccine candidates are presently in various stages of clinical trials, both as prophylactic vaccines for subjects not yet exposed to the virus and as therapeutic vaccines in an attempt to alter the disease course in patients already infected. The cellular immune response to these vaccine candidates has begun to be investigated in a number of these trials. The majority of these vaccine candidates are based on recombinant HIV-1 proteins and would not be expected to generate significant class I-restricted CTL responses, but have induced class II-restricted, $CD4^+$, envelope-specific CTL. Following repeated immunization with the MicroGeneSys gp160 immunogen in an alum, adjuvant CD4-positive, class II-restricted CTL were identified in three of eight recipients (ORENTAS et al. 1990), and an epitope in a highly conserved region of gp41 (aa 584–595) was identified as major target for this response (HAMMOND et al. 1991). Recently class II-restricted CTL clones specific for gp120 which are able to lyse infected cells have also been identified in recipients of this vaccine (SILICIANO et al. 1992).

Recent studies using a primary immunization with a recombinant HIV-1 envelope-vaccinia virus followed by boosting with recombinant gp160 vaccine have yielded evidence that vaccination can also induce class I-restricted CTL, which can be detected following in vitro stimulation with viral antigen (HAMMOND et al. 1992; SILICIANO et al. 1992; COONEY et al. 1993). The CTL generated by this vaccination regimen consist of strain-specific CTL directed at a gp120 epitope as well as more broadly reactive CTL specific for a relatively conserved region of gp41 (R.P.J., S. Hammond, R. Siliciano, and B.D.W., manuscript in preparation). A similar prime-boost regimen in a separate cohort of subjects also demonstrated striking induction of HIV-1-specific CTL activity, including both $CD4^+$ and $CD8^+$ CTL responses (COONEY et al. 1993).

Relatively few data are available regarding the CTL response induced by therapeutic vaccination of seropositive subjects. CTL responses to HIV-1 envelope protein have been investigated in recipients of the MicroGeneSys soluble gp160 vaccine, which was associated with variable increases in both $CD4^+$, class II-restricted as well as $CD8^+$, class I-restricted CTL (KUNDU et al. 1992). Confirmation of therapeutic vaccine-induced CTL responses as well as correlation of immunization with improved clinical and virologic parameters are still needed.

5 Conclusions

Data from numerous investigators have now confirmed that HIV-1 infection is associated with an extremely vigorous CTL response, which is broadly directed at multiple epitopes. Multiple restricting class I antigens have been identified, and the functional activities of the HIV-1-specific CTL have begun to be defined. Nevertheless, the precise role of the CTL response as it relates to the pathogenesis of AIDS remains unclear.

By analogy to other animal models of viral disease, it is likely that CTL directed against HIV-1 play a significant role as a host defense in this infection. Further definition of the importance of the HIV-specific CTL response is likely to come from a number of different avenues of research. A protective immune response to a primate retrovirus appears to have been achieved in the rhesus macaque model, in which protection against high dose challenge has been demonstrated using a live attenuated SIV which is deleted of the nef protein (DANIEL et al. 1992). This model offers the opportunity to define the contribution of virus-specific CTL to the observed protection. The role of CTL in containing virus replication and disease progression in long-term seropositive persons who maintain normal CD4 counts is also likely to contribute to our understanding of the protective role of the cellular immune response in this infection. In this regard, more data are needed to determine whether relative protective effects are conferred by recognition of specific CTL epitopes, whether the CTL response can be broadened by specific immunization strategies, and how immune escape contributes to pathogenesis. Although target cells expressing recombinant viral antigens are readily lysed, more data are needed regarding the ability of these CTL to lyse infected target cells, including infected cells other than $CD4^+$ lymphocytes. Additional studies are needed to clarify the mechanism of declining CTL responses with disease progression and to clarify the potential pathogenic role of a persistent and vigorous CTL response, such as by direct tissue injury or through release of inflammatory cytokines. Further studies are also needed to clarify the potential role of $CD4^+$, class II-restricted CTL in this infection. Answers to these and other questions posed by the continuing and expanding HIV epidemic should facilitate the development of both therapeutic and preventive interventions and should greatly enhance understanding of the role of cellular immunity in human viral diseases in general.

References

Achour A, Picard O, Zagury D, Sarin PS, Gallo RC, Naylor PH, Goldstein AL (1990) HGP-30, a synthetic analogue of human immunodeficiency virus (HIV) p17, is a target for cytotoxic lymphocytes in HIV-infected individuals. Proc Natl Acad Sci USA 87: 7045–7049

Albert JB, Abrahamsson E, Nagy E, Aurelius H, Grines G, Mystrom G, Fenyo EM (1990) Rapid development of isolate-specific neutralizing antibodies after primary HIV-1 infection and consequent emergence of virus variants which resist neutralization by autologous sera. AIDS 4: 107–112

Amiel C, Béné MC, May T, Canton P, Faure GC (1988) LFA-1 expression in HIV infection. AIDS 2: 211–214

Autran B, Mayaud CM, Raphael M, Plata F, Denis M, Bourguin A, Guillon JM, Debre P, Akoun G (1988) Evidence for a cytotoxic T-lymphocyte alveolitis in human immunodeficiency virus-infected patients. AIDS 2: 179–183

Autran B, Plata F, Guillon JM, Joly P, Mayaud C, Debre P (1990) HIV-specific cytotoxic T lymphocytes directed against alveolar macrophages in HIV-infected patients. Res Virol 141: 131–136

Autran B, Sadat SB, Hadida F, Parrot A, Guillon JM, Plata F, Mayaud C, Debre P (1991) HIV-specific cytotoxic T lymphocytes against alveolar macrophages: specificities and down-regulation. Res Virol 142: 113–118

Barre-Sinoussi F, Chermann JC, Rey R, Nugeyre MT, Chamaret S, Gruest J, Daguet C, Axler-Blin C, Vezinet-Brun F, Rouziousx C, Rozenbaum W, Montagnier L (1983) Isolation of a T-lymphotrophic retrovirus from a patient at risk for acquired immunodeficiency syndrome (AIDS). Science 220: 868–871

Bennink JR, Yewdell JW (1988) Murine cytotoxic T lymphocyte recognition of individual influenza virus proteins: high frequency of nonresponder MHC class I alleles. J Exp Med 168: 1935–1939

Bennink JR, Yewdell JW, Gerhard W (1982) A viral polymerase involved in recognition of influenza virus-infected cells by a cytotoxic T-cell clone. Nature 296: 75–76

Biswas P, Poli G, Kinter AL, Justement JS, Stanley SK, Maury WJ, Bressler P, Orenstein JM, Fauci AS (1992) Interferon gamma induces the expression of human immunodeficiency virus in persistently infected promonocytic cells (U1) and redirects the production of virions to intracytoplasmic vacuoles in phorbol myristate acetate-diferentiated U1 cells. J Exp Med 176: 739–750

Bourgault I, Venet A, Levy JP (1992) Three epitopic peptides of the simian immunodeficiency virus Nef protein recognized by macaque cytolytic T lymphocytes. J Virol 66: 750–756

Braciale TJ, Sweetser MT, Morrison LA, Kittlesen DJ, Braciale VL (1989). Class I major histocompatibility complex-restricted cytolytic T lymphocytes recognize a limited number of sites on the influenza hemagglutinin. Proc Natl Acad Sci USA 86: 277–281

Buseyne F, Griscilli C, Blanche S, Schmitt D, Rivière Y (1993a) Detection of HIV-specific cell-mediated cytotoxicity in the peripheral blood from infected children. J Immunol 150: 3569–3581

Buseyne F, McChesney M, Porrot F, Kovarik S, Guy B, Rivière Y (1993b) Gag-specific cytotoxic T lymphocytes from HIV-1 infected individuals: gag epitopes are clustered in three regions of the p24 gag protein. J Virol 67: 694–702

Callahan KM, Fort MM, Obah EA, Reinherz EL, Siliciano RF (1990) Genetic variability in HIV-1 gp120 affects interactions with HLA molecules and T cell receptor. J Immunol 144: 3341–3346

Chen WC, Shen L, Miller MD, Ghim SH, Hughes AL, Letvin NL (1992) Cytotoxic T lymphocytes do not appear to select for mutations in an immunodominant epitope of simian immunodeficiency virus gag. J Immunol 149: 4060–4066

Chicz RM, Urban RG, Lane WS, Gorga JC, Stern LJ, Vignali DA, Strominger JL (1992) Predominant naturally processed peptides bound to HLA-DR1 are derived from MHC-related molecules and are heterogeneous in size. Nature 358: 764–768

Choppin J, Martinon F, Gomard E, Bahraoui E, Connan F, Bouillot M, Levy JP (1990) Analysis of physical interactions between peptides and HLA molecules and application to the detection of human immunodeficiency virus 1 antigenic peptides. J Exp Med 172: 889–899

Choppin J, Martinon F, Connan F, Gomard E, Levy J-P (1991a) HLA-binding regions of HIV-1 proteins. I. Detection of seven HLA binding regions in the HIV-1 nef protein. J Immunol 147: 569–574

Choppin J, Martinon F, Connan F, Pauchard M, Gomard E, Levy JP (1991b) HLA-binding regions of HIV-1 proteins. II. A systematic study of viral proteins. J Immunol 147: 575–583

Claverie JM, Kourilsky P, Langlade DP, Chalufour PA, Dadaglio G, Tekaia F, Plata F, Bougueleret L (1988) T-immunogenic peptides are constituted of rare sequence patterns. Use in the identification of T epitopes in the human immunodeficiency virus gag protein. Eur J Immunol 18: 1547–1553

Clerici M, Lucey DR, Zajac RA, Boswell RN, Gebel HM, Takahashi H, Berzofsky JA, Shearer GM (1991) Detection of cytotoxic T lymphocytes specific for synthetic peptides of gp160 in HIV-seropositive individuals. J Immunol 146: 2214–2219

Cooney EL, McElrath MJ, Corey L, Hu SL, Collier AC, Arditti D, Hoffman M, Coombs RW, Smith GE, Greenberg PD (1993) Enhanced immunity to human immunodeficiency virus (HIV) envelope elicited by a combined vaccine regimen consisting of priming with a vaccinia recombinant expressing HIV envelope and boosting with gp160 protein. Proc Natl Acad Sci USA 90: 1882–1886

Culmann B, Gomard E, Kieny MP, Guy B, Dreyfus F, Saimot AG, Sereni D, Levy JP (1989) An antigenic peptide of the HIV-1 NEF protein recognized by cytotoxic T lymphocytes of seropositive individuals in association with different HLA-B molecules. Eur J Immunol 19: 2383–2386

Culmann B, Gomard E, Kieny MP, Guy B, Dreyfus F, Saimot AG, Sereni D, Sicard D, Levy JP (1991) Six epitopes reacting with human cytotoxic $CD8^+$ T cells in the central region of the HIV-1 NEF protein. J Immunol 146: 1560–1565

Curiel TC, Wong JW, Gorcyzka P, Schooley RT, Walker BD (1993) $CD4^+$ HIV-1 envelope-specific cytotoxic T-lymphocytes derived from the peripheral blood cells of an HIV-1 infected individual. AIDS Res Hum Retroviruses 9: 61–68

Dadaglio G, Leroux A, Langlade DP, Bahraoui EM, Traincard F, Fisher R, Plata F (1991) Epitope recognition of conserved HIV envelope sequences by human cytotoxic T lymphocytes. J Immunol 147: 2302–2309

Dai LC, West K, Littaua R, Takahashi K, Ennis FA (1992) Mutation of human immunodeficiency virus type 1 at amino acid 585 on gp41 results in loss of killing by $CD8^+$ A24-restricted cytotoxic T lymphocytes. J Virol 66: 3151–3154

Daniel M, Kirchoff F, Czajak SC, Sehgal PK, Desrosiers RC (1992) Protective effects of a live attenuated vaccine with a deletion in the nef gene. Science 258: 1938–1941

Del Val M, Münch K, Reddehase MJ, Koszinowski UH (1989) Presentation of CMV immediate-early antigen to cytolytic T lymphocytes is selectively prevented by viral genes expressed in the early phase. Cell 58: 305–313

Falk K, Rotzschke O, Deres K, Metzger J, Jung G, Rammensee HG (1991a). Identification of naturally processed viral nonapeptides allows their quantification in infected cells and suggests an allele-specific T cell epitope forecast. J Exp Med 174: 425–434

Falk K, Rotzschke O, Stevanovic S, Jung G, Rammensee HG (1991b) Allele-specific motifs revealed by sequencing of self-peptides eluted from MHC molecules. Nature 351: 290–296

Frelinger JA, Gotch FM, Zweerink H, Wain E, McMichael AJ (1990) Evidence of widespread binding of HLA class I molecules to peptides. J Exp Med 172: 827–834

Gallo RC, Salahuddin SZ, Popovic M, Shearer GM, Kaplan M, Haynes BF, Palker TJ, Redfield R, Oleske J, Safai B, White G, Foster P, Markham PD (1984) Frequent detection and isolation of cytopathic retroviruses (HTLV-III) from patients with AIDS and at risk for AIDS. Science 224: 500–502

Gotch FM, Nixon DF, Alp N, McMichael AJ, Borysiewicz LK (1990) High frequency of memory and effector gag specific cytotoxic T lymphocytes in HIV seropositive individuals. Int Immunol 2: 707–712

Hadida F, Parrot A, Kieny MP, Sadat SB, Mayaud C, Debre P, Autran B (1992) Carboxyl-terminal and central regions of human immunodeficiency virus-1 NEF recognized by cytotoxic T lymphocytes from lymphoid organs. An in vitro limiting dilution analysis. J Clin Invest 89: 53–60

Hammond SA, Obah E, Stanhope P, Monell CR, Strand M, Robbins FM, Bias WB, Karr RW, Koenig S, Siliciano RF (1991) Characterization of a conserved T cell epitope in HIV-1gp41 recognized by vaccine-induced human cytolytic T cells. J Immunol 146: 1470–1477

Hammond SA, Bollinger RC, Stanhope PE, Quinn TC, Schwartz D, Clements ML, Siliciano RF (1992) Comparative clonal analysis of human immunodeficiency type 1 (HIV-1)-specific $CD4^+$ and $CD8^+$ cytotoxic T lymphocytes isolated from seronegative humans immunized with candidate HIV-1 vaccines. J Exp Med 176: 1531–1543

Harrer T, Harrer E, Jassoy T, Johnson R, Walker B (1993) Induction of HIV-1 replication in a chronically infected T cell line by cytotoxic T lymphocytes. J Acquir Immune Defic Syndr 6: 865–871

Hartshorn KL, Neumeyer D, Vogt MW, Schooley RT, Hirsch MS (1987) Activity of interferons alpha, beta, and gamma against human immunodeficiency virus replication in vitro. AIDS Res Hum Retroviruses 3: 125–133

Heyes M, Brew B, Martin A, Price R, Salazar A, Sidtis J, Yergey J, Mouradian M, Sadler A, Keilp J, Rubinow D, Markey S (1991) Quinolonic acid in cerebrospinal fluid and serum in HIV-1 infection: relationship to clinical and neurological status. Ann Neurol 29: 202–209

Ho M, Armstrong J, Mc Mahon D, Pazin G, Huang XL, Rinaldo C, White-side T, Tripoli C, Levine G, Moody D et al. (1993) A phase 1 study of adoptive transfer of autologous $CD8^+$ T lymphocytes in patients with acquired immunodeficiency syndrome (AIDS)-related complex or AIDS. Blood 81: 2093–2101

Hoffenbach A, Langlade DP, Dadaglio G, Vilmer E, Michel F, Mayaud C, Autran B, Plata F (1989) Unusually high frequencies of HIV-specific cytotoxic T lymphocytes in humans. J Immunol 142: 452–462

Hosmalin A, Clerici M, Houghten R, Pendleton CD, Flexner C, Lucey DR, Moss B, Germain RN, Shearer GM, Berzofsky JA (1990) An epitope in human immunodeficiency virus 1 reverse transcriptase recognized by both mouse and human cytotoxic T lymphocytes. Proc Natl Acad Sci USA 87: 2344–2348

Jardetzky TS, Lane WS, Robinson RA, Madden DR, Wiley DC (1991) Identification of self peptides bound to purified HLA-B27. Nature 353: 326–329

Jassoy C, Johnson RP, Navia BA, Worth J, Walker BD (1992) Detection of a vigorous HIV-1-specific cytotoxic T lymphocyte response in cerebrospinal fluid from infected persons with AIDS dementia complex. J Immunol 149: 3113–3119

Jassoy CJ, Harrer T, Rosenthal T, Navia BA, Worth J, Johnson RP, Walker BD (1993) HIV-1-specific cytotoxic T cells release interferon gamma, tumor necrosis factor (TNF)-alpha and TNF-beta when they encounter their target antigens. J Virol 67: 2844–2852

Johnson RP, Trocha A, Yang L, Mazzara GP, Panicali DL, Buchanan TM, Walker BD (1991) HIV-1 gag-specific cytotoxic T lymphocytes recognize multiple highly conserved epitopes. Fine specificity of the gag-specific response defined by using unstimulated peripheral blood mononuclear cells and cloned effector cells. J Immunol 147: 1512–1521

Johnson RP, Trocha A, Buchanan TM, Walker BD (1992) Identification of overlapping HLA class I-restricted cytotoxic T cell epitopes in a conserved region of the human immunodeficiency virus type 1 envelope glycoprotein: definition of minimum epitopes and analysis of the effects of sequence variation. J Exp Med 175: 961–971

Johnson RP, Trocha A, Buchanan T, Walker BD (1993) Recognition of a highly conserved region of human immunodeficiency virus type I gp120 by a HLA-Cw4-restricted cytotoxic T lymphocyte clone. J Virol 67: 438–445

Joly E, Mucke L, Oldstone MB (1991) Viral persistence in neurons explained by lack of major histocompatibility class I expression. Science 253: 1283–1285

Kerkau T, Schmitt LR, Schimpl A, Wecker E (1989) Downregulation of HLA class I antigens in HIV-1-infected cells. AIDS Res Hum Retroviruses 5: 613–620

Koenig S, Earl P, Powell D, Pantaleo G, Merli S, Moss B, Fauci AS (1988) Group-specific, major histocompatibility complex class I-restricted cytotoxic responses to human immunodeficiency virus 1 (HIV-1) envelope proteins by cloned peripheral blood T cells from an HIV-1-infected individual. Proc Natl Acad Sci USA 85: 8638–8642

Koenig S, Fuerst S, Wood LV, Woods R, Jones G, Fauci AS (1990a) Structural and functional characteristics of HIV-specific CTL. 6th international conference on AIDS, San Francisco, CA

Koenig S, Fuerst TR, Wood LV, Woods RM, Suzich JA, Jones GM, de ICVF, Davey RTJ, Venkatesan S, Moss B et al. (1990b) Mapping the fine specificity of a cytolytic T cell response to HIV-1 nef protein. J Immunol 145: 127–135

Koenig S, Jones G, Boone E, Brewah A, Newell A, Torbett B, Mosier D, Fauci AS (1991) Inhibition of HIV replication by HIV-specific CTL. 7th international conference on AIDS, Florence, Italy

Koenig S, van Kuyk R, Jones G, Torbett B, Mosier D, Brewah Y, Leath S, Davey V, Yannelli J, Rosenberg S, Fauci A, Lane H (1992) Adoptive transfer of a nef-specific CTL clone into hu-PBL-SCID mice and an HIV-infected patient. 8th international conference on AIDS/III STD World Congress, Amsterdam, The Netherlands, ThA1573

Koup RA, Sullivan JL, Levine PH, Brettler D, Mahr A, Mazzara G, McKenzie S, Panicali D (1989) Detection of major histocompatibility complex class I-restricted, HIV-specific cytotoxic T lymphocytes in the blood of infected hemophiliacs. Blood 73: 1909–1914

Koup R, Hesselton R, Johnson RP, Walker BD, Sullivan J (1991a) Effect of adoptively transferred CTL on HIV replication in the Hu-PBL-SCID mouse model. National Cooperative Vaccine Development Conference, Marco Island, FL

Koup RA, Pikora CA, Luzuriaga K, Brettler DB, Day ES, Mazzara GP, Sullivan JL (1991b) Limiting dilution analysis of cytotoxic T lymphocytes to human immunodeficiency virus gag antigens in infected persons: in vitro quantitation of effector cell populations with p17 and p24 specificities. J Exp Med 174: 1593–1600

Kundu S, Katzenstein D, Moses L, Merigan T (1992) Enhancement of human immunodeficiency virus (HIV) specific $CD4^+$ and $CD8^+$ cytotoxic T-lymphocyte activities in HIV infected asymptomatic patients given recombinant gp160 vaccine. Proc Natl Acad Sci USA 89: 11204–11208

Kundu SK, Merigan TC (1992) Equivalent recognition of HIV proteins, Env, Gag and Pol, by $CD4^+$ and $CD8^+$ cytotoxic T-lymphocytes. AIDS 6: 643–649

Lieberman J (1993) ACTG, Washington, D.C.

Lieberman J, Fabry JA, Kuo MC, Earl P, Moss B, Skolnik PR (1992) Cytotoxic T lymphocytes from HIV-1 seropositive individuals recognize immunodominant epitopes in gp160 and reverse transcriptase. J Immunol 148: 2738–2747

Littaua RA, Oldstone MB, Takeda A, Debouck C, Wong JT, Tuazon CU, Moss B, Kievits F, Ennis FA (1991) An HLA-C-restricted $CD8^+$ cytotoxic T-lymphocyte clone recognizes a highly conserved epitope on human immunodeficiency virus type 1 gag. J Virol 65: 4051–4056

Littaua RA, Oldstone MB, Takeda A, Ennis FA (1992) A $CD4^+$ cytotoxic T-lymphocyte clone to a conserved epitope on human immunodeficiency virus type 1 p24: cytotoxic activity and secretion of interleukin-2 and interleukin-6. J Virol 66: 608–611

Luzuriaga K, Koup RA, Pikora CA, Brettler DB, Sullivan JL (1991) Deficient human immunodeficiency virus type 1-specific cytotoxic T cell responses in vertically infected children. J Pediatr 119: 230–236.

Matsuyama T, Kobayashi N, Yamamoto N (1991) Cytokines and HIV infection: is AIDS a tumor necrosis factor disease? AIDS 5: 1405–1417

McChesney M, Tanneau F, Regnault A, Sansonetti P, Montagnier L, Kieny MP, Rivière Y (1990) Detection of primary cytotoxic T lymphocytes specific for the envelope glycoprotein of HIV-1 by deletion of the env amino-terminal signal sequence. Eur J Immunol 20: 215–220

Meyerhans A, Dadaglio G, Vartanian JP, Langlade DP, Frank R, Asjo B, Plata F,S. W-H (1991) In vivo persistence of a HIV-1-encoded HLA-B-27-restricted cytotoxic T lymphocyte epitope despite specific in vitro reactivity. Eur J Immunol 21: 2637–2640

Miller MD, Lord CI, Stallard V, Mazzara GP, Letvin NL (1990) The gag-specific cytotoxic T lymphocytes in rhesus monkeys infected with the simian immunodeficiency virus of macaques. J Immunol 144: 122–128

Miller MD, Yamamoto H, Hughes AL, Watkins DI, Letvin NL (1991) Definition of an epitope and MHC class I molecule recognized by gag-specific cytotoxic T lymphocytes in SIVmac-infected rhesus monkeys. J Immunol 147: 320–329

Myers G, Berzofsky JA, Rabson AB, Smith TF, Wong-Staal F (eds) (1991) Human retroviruses and AIDS 1991: a compilation and analysis of nucleic acid and amino acid sequences. Los Alamos, N.M., Los Alamos National Laboratory

Nara PL, Smit L, Dunlop N, Hatch W, Merges M, Waters D, Kelliher J, Gallo RC, Fischinger PJ, Goudsmit J (1990) Emergence of viruses resistant to neutralization by V3-specific antibodies in experimental human immunodeficiency virus type 1 IIIB infection of chimpanzees. J Virol 64: 3779–3791

Nixon DF, McMichael AJ (1991) Cytotoxic T-cell recognition of HIV proteins and peptides [editorial]. AIDS 5: 1049–1059

Nixon DF, Townsend AR, Elvin JG, Rizza CR, Gallwey J, McMichael AJ (1988) HIV-1 gag-specific cytotoxic T lymphocytes defined with recombinant vaccinia virus and synthetic peptides. Nature 336: 484–487

Nixon DF, Huet S, Rothbard J, Kieny MP, Delchambre M, Thiriart C, Rizza CR, Gotch FM,McMichael AJ (1990) An HIV-1 and HIV-2 cross-reactive cytotoxic T-cell epitope. AIDS 4: 841–845

Orentas RJ, Hildreth JE, Obah E, Polydefkis M, Smith GE, Clements ML, Siliciano RF (1990). Induction of $CD4^+$ human cytolytic T cells specific for HIV-infected cells by a gp160 subunit vaccine. Science 248: 1234–1237

Phillips RE, Rowland JS, Nixon DF, Gotch FM, Edwards JP, Ogunlesi AO, Elvin JG, Rothbard JA, Bangham CR, Rizza CR et al. (1991) Human immunodeficiency virus genetic variation that can escape cytotoxic T cell recognition. Nature 354: 453–459

Plata F, Autran B, Martins LP, Wain HS, Raphael M, Mayaud C, Denis M, Guillon JM, Debre P (1987) AIDS virus-specific cytotoxic T lymphocytes in lung disorders. Nature 328: 348–351

Reimann KA, Snyder GB, Chalifoux LV, Waite BC, Miller MD, Yamamoto H, Spertini O, Letvin NL (1991) An activated $CD8^+$ lymphocyte appears in lymph nodes of rhesus monkeys early after infection with simian immunodeficiency virus. J Clin Invest 88: 1113–1120

Reitz MS Jr., Wilson C, Naugle C, Gallo RC, Robert-Guroff M (1988) Generation of a neutralization-resistant variant of HIV-1 is due to selection for a point mutation in the envelope gene. Cell 54: 57–63

Rivière Y, Tanneau SF, Regnault A, Lopez O, Sansonetti P, Guy B, Kieny MP, Fournel JJ, Montagnier L (1989a) Human immunnodeficiency virus-specific cytotoxic responses of seropositive individuals: distinct types of effector cells mediate killing of targets expressing gag and env proteins. J Virol 63: 2270–2277

Rivière Y, Tanneau SF, Regnault A, Lopez O, Sansonetti P, Guy B, Kieny MP, Fournel JJ, Montagnier L (1989b) Multiple cytotoxic effector cells are induced by infection with the human immunodeficiency virus. Res Immunol 140: 110–115

Rotzschke O, Falk K, Deres K, Schild H, Norda M, Metzger J, Jung G, Rammensee HG (1990) Isolation and analysis of naturally processed viral peptides as recognized by cytotoxic T cells. Nature 348: 252–254

Rudensky A, Preston HP, Hong SC, Barlow A, Janeway CAJ (1991) Sequence analysis of peptides bound to MHC class II molecules. Nature 353: 622–627

Scheppler JA, Nicholson JK, Swan DC, Ahmed AA, McDougal JS (1989) Down-modulation of MHC-I in a $CD4^+$ T cell line, CEM-E5, after HIV-1 infection. J Immunol 143: 2858–2866

Selmaj KW, Raine CS (1988) Tumor necrosis factor mediates myelin and oligodendrocyte damage in vitro. Ann Neurol 23: 339–346

Selmaj KW, Farooq M, Norton WT, Raine CS, Brosnan CF (1990) Proliferation of astrocytes in vitro in response to cytokines. A primary role for tumor necrosis factor. J Immunol 144: 129–135

Sethi KK, Naher H, Stroehmann I (1988) Phenotypic heterogeneity of cerebrospinal fluid-derived HIV-specific and HLA-restricted cytotoxic T-cell clones. Nature 335: 171–181

Sethna M, Lampson L (1991) Immune modulation within the brain: recruitment of inflammatory cells and increased major histocompatibility antigen expressing following intracerebral injection of interferon-gamma. J Neuroimmunol 34: 121–132
Shen L, Chen ZW, Miller MD, Stallard V, Mazzara GP, Panicali DL, Letvin NL (1991) Recombinant virus vaccine-induced SIV-specific CD8+ cytotoxic T lymphocytes. Science 252: 440–443
Siliciano RF, Lawton T, Knall C, Karr RW, Berman P, Gregory T, Reinherz EL (1988) Analysis of host-virus interactions in AIDS with anti-gp 120 T cell clones: effect of HIV sequence variation and a mechanism for CD4+ cell depletion. Cell 54: 561–575
Siliciano RF, Knall C, Lawton T, Berman P, Gregory T, Reinherz EL (1989) Recognition of HIV glycoprotein gp120 by T cells. Role of monocyte CD4 in the presentation of gp120. J Immunol 142: 1506–1511
Siliciano RF, Bollinger RC, Callahan KC, Hammond SA, Liu AY, Miskovsky EP, Rowell JF, Stanhope PE (1992) Clonal analysis of T-cell responses to the HIV-1 envelope proteins in AIDS vaccine recipients. AIDS Res Human Retroviruses 8: 1349–1352
Takahashi H, Cohen J, Hosmalin A, Cease KB, Houghten R, Cornette JL, DeLisi C, Moss B, Germain RN, Berzofsky JA (1988) An immunodominant epitope of the human immunodeficiency virus envelope glycoprotein gp160 recognized by class I major histocompatibility complex molecule-restricted murine cytotoxic T lymphocytes. Proc Natl Acad Sci USA 85: 3105–3109
Takahashi K, Dai LC, Fuerst TR, Biddison WE, Earl PL, Moss B, Ennis FA (1991) Specific lysis of human immunodeficiency virus type 1-infected cells by a HLA-A3.1-restricted CD8+ cytotoxic T-lymphocyte clone that recognizes a conserved peptide sequence within the gp41 subunit of the envelope protein. Proc Natl Acad Sci USA 88: 10277–10281
Townsend A, Bodmer H (1989) Antigen recognition by class I-restricted T lymphocytes. Annu Rev Immunol 7: 601–624
Townsend AR, McMichael AJ, Carter NP, Huddleston JA, Brownlee GG (1984) Cytotoxic T-cell recognition of the influenza nucleoprotein and hemagglutinin expressed in transfected mouse L cells. Cell 39: 13–25
Townsend AR, Bastin J, Gould K, Brownlee GG (1986). Cytotoxic T lymphocytes recognize influenza haemagglutinin that lacks a signal sequence. Nature 324: 575–577
Tsomides TJ, Walker BD, Eisen HN (1991) An optimal peptide recognized by CD8+ T cells binds very tightly to the restricting class I major histocompatibility complex protein on intact cells but not to the purified class I protein. Proc Natl Acad Sci USA 88: 11276–11280
Tsubota H, Lord CI, Watkins DI, Morimoto C, Letvin NL (1989) A cytotoxic T lymphocyte inhibits acquired immunodeficiency syndrome virus replication in peripheral blood lymphocytes. J Exp Med 169: 1421–1434
Tyler DS, Nastala CL, Stanley SD, Matthews TJ, Lyerly HK, Bolognesi DP, Weinhold KJ (1989) GP120 specific cellular cytotoxicity in HIV-1 seropositive individuals. Evidence for circulating CD16+ effector cells armed in vivo with cytophilic antibody. J Immunol 142: 1177–1182
Venet A, Walker BD (1993) Epitopes in HIV/SIV infection. AIDS 7 (Suppl 1): S117–S126
Venet A, Bourgault I, Aubertin AM, Kieny MP, Levy JP (1992) Cytotoxic T lymphocyte response against multiple simian immunodeficiency virus A (SIV) proteins in SIV-infected macaques. J Immunol 148: 2899–2908
Vidovic M, Sparacio S, Elovitz M, Benveniste E (1990) Induction and regulation of class II major histocompatibility complex mRNA expression in astrocytes by interferon-gamma and tumor necrosis factor-alpha. J Neuroimmunol 30: 189–200
Vowels BR, Gershwin ME, Gardner MB, Ahmed AA, McGraw TP (1989) Characterization of simian immunodeficiency virus-specific T-cell-mediated cytotoxic response of infected rhesus macaques. AIDS 3: 785–792
Walker BD, Chakrabarti S, Moss B, Paradis TJ, Flynn T, Durno AG, Blumberg RS, Kaplan JC, Hirsch MS, Schooley RT (1987). HIV-specific cytotoxic T lymphocytes in seropositive individuals. Nature 328: 345–348
Walker BD, Flexner C, Paradis TJ, Fuller TC, Hirsch MS, Schooley RT, Moss B (1988) HIV-1 reverse transcriptase is a target for cytotoxic T lymphocytes in infected individuals. Science 240: 64–66
Walker BD, Flexner C, Birch LK, Fisher L, Paradis TJ, Aldovini A, Young R, Moss B, Schooley RT (1989) Long-term culture and fine specificity of human cytotoxic T-lymphocyte clones reactive with human immunodeficiency virus type 1. Proc Natl Acad Sci USA 86: 9514–9518

Walker CM, Levy JA (1989) A diffusible lymphokine produced by $CD8^+$ T lymphocytes suppresses HIV replication. Immunology 66: 628–630
Walker CM, Moody DJ, Stites DP, Levy JA (1986) $CD8^+$ lymphocytes can control HIV infection in vitro by suppressing virus replication. Science 234: 1563–1566
Warner JF, Anderson CG, Laube L, Jolly DJ, Townsend K, Chada S, St Louis D (1991) Induction of HIV-specific CTL and antibody responses in mice using retroviral vector-transduced cells. Aids Res Hum Retroviruses 7: 645–655
Weinhold KJ, Lyerly HK, Matthews TJ, Tyler DS, Ahearne PM, Stine KC, Langlois AJ, Durack DT, Bolognesi DP (1988) Cellular anti-GP120 cytolytic reactivities in HIV-1 seropositive individuals. Lancet 1: 902–905
Werner E, Bitterlich G, Fuchs D, Hausen A, Reibnegger G, Szabo G, Dierich M, Wachter H (1987) Human macrophages degrade tryptophan upon induction by interferon-gamma. Life Sci 41: 273–280
Yamamoto H, Miller MD, Tsubota H, Watkins DI, Mazzara GP, Stallard V, Panicali DL, Aldovini A, Young RA, Letvin NL (1990a) Studies of cloned simian immunodeficiency virus-specific T lymphocytes. Gag-specific cytotoxic T lymphocytes exhibit a restricted epitope specificity. J Immunol 144: 3385–3391
Yamamoto H, Miller MD, Watkins DI, Snyder GB, Chase NE, Mazzara GP, Gritz L, Panicali DL, Letvin NL (1990b) Two distinct lymphocyte populations mediate simian immunodeficiency virus envelope-specific target cell lysis. J Immunol 145: 3740–3746
Yamamoto H, Ringler DJ, Miller MD, Yasutomi Y, Hasunuma T, Letvin NL (1992) Simian immunodeficiency virus-specific cytotoxic T lymphocytes are present in the AIDS-associated skin rash in rhesus monkeys. J Immunol 149: 728–734
Yewdell JW, Bennink JR (1992) Cell biology of antigen processing and presentation to major histocompatibility complex class I molecule-restricted T lymphocytes. Adv Immunol 52: 1–123
Yewdell JW, Bennink JR, Smith GL, Moss B (1985) Influenza A virus nucleoprotein is a major target antigen for cross-reactive anti-influenza A virus cytotoxic T lymphocytes. Proc Natl Acad Sci USA 82: 1785–1789

Cytotoxic T Lymphocytes in Human Immunodeficiency Virus Infection: Regulator Genes

Y. RIVIERE, M. N. ROBERTSON, and F. BUSEYNE

1 Introduction

Like all retroviruses, the human immunodeficiency virus (HIV) provirus contains two long terminal repeat elements (LTR) along with the three structural genes—*gag, pol,* and *env*—that are essential for virus replication. In addition to these genetics elements, HIV contains at least six additional genes: *tat, rev, nef, vif, vpr,* and *vpu,* whose functions impart both positive and negative effects on the HIV life cycle (for review see CULLEN 1991; GREENE 1991; HASELTINE 1991; STEFFIE and WONG STAAL 1991; FEINBERG and GREENE 1992).

After attachment of the virion to its cellular receptor—the CD4 molecule—and penetration of the virion core into the cell, reverse transcription of the RNA genome into double-stranded DNA occurs. Following DNA synthesis, the viral DNA transits to the nucleus, and integration of the viral DNA into cellular DNA occurs to form the provirus. Once the provirus has been incorporated into the cellular DNA, both cellular and viral factors are required to initiate expression of viral genes. After activation of the provirus, the HIV life cycle can be divided into early and late stages by a transition in the character of viral transcripts produced following infection: early viral transcripts are the 2-kb RNA encoding the early regulatory nonstructural proteins with regulatory functions, the Tat, Rev and Nef proteins. The Tat protein is a powerful transactivator of HIV gene expression, and the Rev protein's main

Unité de Virologie et Immunologie Cellulaire, Institut Pasteur, 75724 Paris Cedex 15, France

Current Topics in Microbiology and Immunology, Vol. 189

function is to effect the transport of unspliced and singly spliced mRNA from the nucleus to the cytoplasm. Both Tat and Rev proteins are essential for HIV replication: loss of Tat function results in the absence of viral RNA and protein expression, and in the absence of Rev, viral structural proteins are not synthesized (in contrast to the Tat and Nef regulatory proteins). The transition from the early to the late stage in the virus cycle is dependent on the Rev protein; the Rev protein may act at a post-transcriptional level by activating cytoplasmic expression of unspliced and singly spliced mRNA encoding the products of the *gag*, *pol*, and *env* genes.

In contrast to the essential role of Tat and Rev in HIV replication, the four other additional proteins, Nef, Vif, Vpr, and Vpu have been called nonessential, because they can be deleted without completely abrogating the ability of the virus to replicate.

The p27-kDa Nef is a myristylated protein produced early in infection, since its mRNA is detectable along with the two regulatory nuclear proteins, Tat and Rev. The Nef protein is cytoplasmic and partly associated with membranes; it is not required for virus replication in vitro, but in vivo experiments with Simian immunodeficiency virus (SIV)-infected macaques indicate that in vivo Nef is important for virus expression and pathogenicity (KESTLER et al. 1991). On the cellular level, CD4 and interleukin 2 (IL-2) are downregulated by Nef expression. While the function of Nef still remains controversial, the Nef protein can possibly be classified among the regulatory proteins of HIV.

The precise functional role of the three other additional genes, *vpu*, *vpr*, and *vif*, in the viral cycle is not clear. The p15-kDa Vpu protein seems to interact with gp160–CD4 complexes in the late Golgi apparatus and to be involved in viral release. The p23-kDa Vif protein presumably plays a role in the late step of the replication cycle (infectious virion assembly) or in the early step (virus attachment and entry). The p18-kDa Vpr protein function is not well defined; the Vpr protein is made late in infection and because this protein is present in the virion, it could be considered as a structural protein.

2 Specificity of the Cellular Immune Response in Human Immunodeficiency Virus Type 1-Infected Donors to Non-structural Viral Proteins

HIV-specific cytotoxic lymphocytes which can specifically lyse target cells expressing HIV antigen have been described by a number of investigators (for review see MILLS et al. 1989; NIXON and MCMICHAEL 1991). In HIV-1-infected individuals, two forms of anti-HIV-specific cytotoxic activities can be detected: lysis mediated by major histocompatibility complex (MHC)-restricted cytotoxic T lymphocytes (CTL) and lysis mediated by antibody-dependent cellular cytotoxicity (ADCC), the latter involving non-MHC-restricted cytolytic activities of effector cell populations bearing HIV-specific cytophilic antibodies that direct their lytic activities. In HIV-1-infected individuals, ADCC has been shown to be mainly directed at the envelope glycoprotein (McCHESNEY et al. 1990; TANNEAU et al. 1990), but any other viral protein shown to be expressed at the surface of infected cells may be a target for ADCC.

A characteristic of HIV infection is the long-lasting detection of HIV-specific cytotoxic activities from fresh peripheral blood mononuclear cells (PBMC) without in vitro stimulation; this is likely to be due to a persistent antigenic stimulation in vivo. CTL activity derived from PBMC is vigorous and directed at most viral structural (Gag, Pol, Env) and nonstructural (Tat, Rev, Nef, Vpu, Vif) proteins (for review see AUTRAN and LETVIN 1991; NIXON and McMICHAEL 1991; VENET et al. 1993).

We and other investigators have been able to detect cytotoxic effector cells in the PBMC of HIV-1-infected individuals directed at most viral proteins, including structural and nonstructural proteins. In this paper, we will review cytotoxic activities specific for the nonstructural proteins. Using recombinant vaccinia viruses that express specific HIV proteins and Epstein-Barr virus (EBV)-transformed B lymphocytes as target cells, we have found that a variable percentage of HIV-1-infected donors have cytotoxic effector cells directed at Nef, Vif, and Rev (Table 1 and BUSEYNE et al. 1992). These virus-specific cytotoxic activities were detected in PBMC isolated from the blood immediately after lymphocyte separation and without in vitro stimulation (RIVIÈRE et al. 1989). We and others have found that nef-specific activities are MHC restricted and linked to $CD8^+$ $CD3^+$ T lymphocytes (KOENIG et al. 1990; BUSEYNE et al. 1992). It is likely that the Rev- and Vif-specific activities are also MHC restricted.

Table 1. Specificity and frequency of the cytotoxic activities from human immunodeficiency type 1 (HIV-1)-seropositive donor peripheral blood mononuclear cells

	Circulating[a]		In vitro restimulated			
			I[c]		II[d]	
	n	%	*n*	%	*n*	%
p27Nef	10/25	40	10/20	50	18/28	64
p23Vif	4/18	22	3/6	50	11/27	41
p14Tat	–[b]		2/6	33	1/26	4
p20Rev	3/13	23	7/16	44	8/22	36
p15Vpu	–[b]		–[b]		–[b]	
p18Vpr	–[b]		–[b]		–[b]	
p55Gag	35/54	65	25/29	86	26/28	93

[a] From BUSEYNE et al. 1992.
[b] There are no published data on circulating cytotoxic activities specific for p14Tat nor for any cytotoxic activities specific for p15 Vpu or p18 Vpr.
[c] From BUSEYNE et al. 1993a and unpublished results.
[d] From LAMHAMEDI-CHERADI et al. 1992.

In HIV-infected subjects, in vitro mitogen stimulation of T lymphocytes from PBMC or other lymphoid organs has been used to establish HIV-specific CTL lines or clones. These mitogen-induced effector cells have been used by a number of investigators to define CTL epitopes located in structural as well as nonstructural HIV proteins. Table 1 summarizes the frequency of recognition of the different nonstructural HIV proteins that have been reported (BUSEYNE et al. 1992; LAMHAMEDI-CHERADI et al. 1992). The p27Nef, p23Vif, and p20Rev proteins were recognized by CTL derived from stimulated PBMC with a similar frequency in the two different studies (range, 50%–64% for p27Nef, 41%–50% for p23Vif, and 36%–44% for p20Rev). The difference in the results for the Tat-specific CTL likely reflects the small number of individuals tested (six) in one study. Nothing has been reported concerning p15Vpu- or p18Vpr-specific CTL activity. Taken together, these results show that it has been possible to generate CTL specific to the nonstructural proteins of HIV (p27Nef, p23Vif, and p20Rev) from about 50% of the donors, although the same CTL lines react against p55Gag for 86% or more of the donors studied. This could be due to the fact that non-structural proteins are smaller and less conserved than the p55Gag.

3 Epitope Mapping of p27Nef: Two Regions of the Protein are Targets for Most of the CTL Activities

The p27nef protein CTL recognition has been extensively studied during recent years by several investigators (CHENCINER et al. 1989; CULMAN et al. 1989, 1991; RIVIÈRE et al. 1989; KOENIG et al. 1990; CHEYNIER et al. 1992; LAMHAMEDI-CHERADI et al. 1992; HADIDA et al. 1992a; BUSEYNE et al. 1993b). In particular, CTL lines or clones were derived by in vitro stimulation with mitogen such as phytohemagglutinin A or by anti-CD3 antibody and subsequent expansion with exogenous IL-2. These cell lines or clones were $CD3^+$ $CD8^+$ T cells. After characterization of the p27Nef CTL activity using autologous B-EBV targets infected with specific recombinant vaccinia viruses, epitope mapping of the protein was performed by using as targets B-EBV autologous cells coated with different peptides spanning the Nef amino acid sequence. Twenty-four different peptides have been shown to define epitopic regions within the p27Nef protein. Eighteen are clustered into two adjacent regions of the central portion of Nef (aa 66–106 for 11 peptides, aa 113–147 for seven) and four are clustered into the C-terminal end of the protein (aa 176–206; BOURGAULT et al. 1992; BUSEYNE et al. 1993b; CULMANN et al. 1989, 1991 and unpublished data; HADIDA et al. 1992a; JASSOY et al. 1992; KOENIG et al. 1990; MICHEL et al. 1992; and Table 2). For 23 of the T cell lines or clones, the MHC-restrictive element has been identified.

Table 2. Published epitopic regions within the p27 Nef protein

Species/Virus	Amino acid number	Amino acid sequence	MHC restriction	Reference
Human/HIV	66–80	VGFPVTPQVPLRPMT	A1	HADIDA et al. 1992a
Human/HIV	73–82	QVPLRPMTYK	A3.1	KOENIG et al. 1990
Human/HIV	73–82	QVPLRPMTYK	A3, B35, A11	CULMANN et al. 1991
Human/HIV	73–82	QVPLRPMTYK	B35 or C4	BUSEYNE et al. 1993b
Human/HIV	73–82	QVPLRPMTYK	A2	Robertson and Rivière, unpublished
Human/HIV	74–81	VPLRPMTY	B35	Culmann et al. unpublished
Human/HIV	74–85	VPLRPMTYKAAV	B35 or C4, B17	Buseyne et al. unpublished
Macaque/SIV	108–123	LRAMTYKLAIDMSHFI	nd	BOURGAULT et al. 1992
Human/HIV	78–93	MTYKAAVDLSHFLKEK	nd	CULMANN et al. 1989
Human/HIV	83–94	AAVDLSHFLKEK	A11	CULMANN et al. 1991
Human/HIV	84–91	AVDLSHFL	Bw62	Culmann et al., unpublished
Human/HIV	84–92	AVDLSHFLE	A11	Culmann et al., unpublished
Human/HIV	86–100	DLSHFLKEKGGLEGL	B35 or C4	BUSEYNE et al. 1993b
Human/HIV	89–97	HFLKEKGGL	B8	Culmann et al., unpublished
Human/HIV	93–106	EKGGLEGLIHSQRR	A1	HADIDA et al. 1992a
Human/HIV	113–128	WIYHTQGYFPDWQNYT	B17 & B37	CULMANN et al. 1989
Human/HIV	113–128	WIYHTQGYFPDWQNYT	A1	HADIDA et al. 1992a
Human/HIV	115–125	YHTQGYFPDW	B17	CULMANN et al. 1991
Human/HIV	116–125	HTQGYFPDW	B17	Culmann et al., unpublished
Human/HIV	117–128	TQGYFPDWQNYT	B17 & B37	CULMANN et al. 1991
Human/HIV	117–124	TQGYFPDW	Bw62	Culmann et al., unpublished
Human/HIV	120–128	YFPDWQNYT	B17 & B37	Culmann et al., unpublished
Human/HIV	120–144	YFPDWQ—TFGWCYK	A24	JASSOY et al. 1992
Macaque/SIV	155–169	DWQDYTSGPGIRYPK	nd	BOURGAULT et al. 1992
Human/HIV	126–138	NYTPGPGVRYPLT	B7	CULMANN et al. 1991
Human/HIV	132–147	GVRYPLTFGWCYKLVP	A1	HADIDA et al. 1992a
Human/HIV	132–147	GVRYPLTFGWCYKLVP	B18	CULMANN et al. 1991
Macaque/SIV	164–178	GIRYPKTFGWLWKLV	nd	BOURGAULT et al. 1992
Human/HIV	134–144	GVRYPLTFGWCYKLVP	B18 et B49	Culmann et al., unpublished
Human/HIV	155–168	EANKGENTSLLHPV	nd	Rivière et al., unpublished
Human/HIV	176–185	PEREVLEWRF	nd	Rivière et al., unpublished
Mouse/RVV nef	175–190	DPEREVLEWRFDSRLA	H2d (IA, IE)	MICHEL et al. 1992
Mouse/RVV nef	182–198	EWRFDSRLAFHHVAREL	H2d (IA, IE)	MICHEL et al. 1992
Human/HIV	182–198	EWRFDSRLAFHHVAREL	A1 & B8	HADIDA et al. 1992a
Human/HIV	188–201	RLAFHHVARELHPE	B35 or C4	BUSEYNE et al. 1993b
Human/HIV	192–206	HHVARELHPEYFKNC	A1	HADIDA et al. 1992a

HIV, human immunodeficiency virus; SIV, simian immunodeficiency virus; RVV, recombinant vaccinia virus; MHC, major histocompatibility complex; nd, not determined.

Altogether, 15 distinct HLA molecules were shown to present 19 different peptides.

Four independent laboratories have reported that six different HLA molecules present peptides within the aa 66–85 region and four different HLA molecules have been reported to present peptides within the aa 84–97 region (Table 2). However, because the synthetic peptides that were used were 10–15 amino acids long, which is longer than the 8- to 9-mer expected length for naturally processed viral peptides (RÖTZCHKE et al. 1990; VAN BLEEK and NATHENSON 1990), it is likely that they included more than one unique T cell epitope; accordingly, a recent study performed by Culmann-Penciolelli et al., shows that peptide 73–82 was presented by the A3 and A11 molecules, peptide 74–81 by B35, and peptide 84–92 by A11 (Culmann-Penciolelli et al., personal communication). Within the aa 176–206 sequence of the C-terminal region, HADIDA et al. (1992b) recently reported that several HLA class I molecules (A1, A2, A24, B5, B8, and B31) efficiently present either identical or distinct peptides.

All these results show that the p27Nef protein is a very common target for CTL recognition in HIV-infected persons. Analysis of the various amino acid sequences published for the p27Nef protein revealed that the central region (aa 62–166) is rather conserved within HIV-1, HIV-2, and SIV isolates (MYERS et al. 1992). In experimentally infected macaques, three peptides from the homologous central region of SIV/nef (aa 108–123, aa 155–169, and aa 164–178) have been shown to include a T cell epitope (BOURGAULT et al. 1992; Table 2); these results could be explained by structural similarities between the nonhuman and the human primate MHC molecules. On the other hand, in splenocytes from mice immunized with recombinant vaccinia virus (RVV) encoding the HIV-1/Lai *nef* gene and stimulated in vitro with the homologous recombinant Nef protein, $CD4^+$ cytotoxic T cell activities restricted by MHC class II molecules of the H-2d haplotype were found directed to two peptides within the C-terminal region (MICHEL et al. 1992).

Both the great diversity of Nef peptides recognized by CTL and their HLA restricting elements show that the p27Nef protein is a widely recognized target for CTL in HIV-1-infected subjects, as is the case for the p24gag protein (BUSEYNE et al. 1993a). Because the *nef* gene is expressed early in the viral cycle through a Rev-independent RNA (ROBERT-GUROFF et al. 1990), it is likely that induction of HIV Nef-specific CTL activity will be beneficial to the infected host, allowing elimination of target cells expressing the Nef protein before release of the progeny virus, as has been shown for target cells acutely infected with vaccinia virus (ZINKERNAGEL and ALTHAGE 1977).

A recent report shows that infection with a live, attenuated SIV with a deletion in the Nef gene protects rhesus macaques against challenge by intravenous inoculation of the homologous live pathogenic SIV (DANIEL et al. 1992). The mechanisms responsible for the protective immunity observed in these experiments are not known. A possible role of CTL has been advanced;

surprisingly, the Nef deletion was in the central region of SIV nef (aa 58 to 118), which is very similar to the central region of HIV-1/Nef, where most Nef-specific CTL epitopes in HIV-infected humans are clustered. Another possibility would be protection by viral interference, but this seems unlikely because Nef is required for maintaining high virus loads during the course of persistent infection in vivo and a very small proportion of cells are thought to be infected by the Nef-deleted SIV (KESTLER et al. 1991).

4 Discussion

Analysis of the antigenic specificity of the CTL response directed to non-structural proteins in HIV infection reveals that almost all proteins could be targets for CTL- mediated lysis. A similar observation could be made for the viral structural proteins. In most viral infections, the functions of antigen-specific T lymphocyte are required for recovery from viral infection, for clearance of virus, and for control of persistent infection. In the case of HIV infection, little information is known about the biological role of the CTL in the course of this long infectious process, in which the disease usually appears after a long incubation period.

The control of HIV replication in $CD4^+$ T cells by autologous $CD8^+$ T cells from the PBMC of HIV-infected donors is a well-established phenomenon (WALKER et al. 1986). The precise mechanism of this antiviral activity, as well as the identification of the CD8 subsets responsible for this activity, needs further investigation. The antiviral activity has been shown to be mediated by a soluble $CD8^+$ T cell-derived factor that has not yet been identified, but the most efficient inhibition of virus replication further requires contact between the infected $CD4^+$ T cells and effector $CD8^+$ T cells (for review see RIVIÈRE 1990). Interestingly, $CD8^+$ T cells from patients with acquired immunodeficiency syndrome (AIDS) showed reduced or no such inhibitory activity (BRINCHMANN et al. 1990). Although the killing process per se may or may not be involved in the mechanism by which $CD8^+$ T cells control HIV replication (WALKER et al. 1991), this does not rule out the possibility that the same subset of $CD8^+$ T cells, i.e., HIV-specific CTL, are responsible for this activity. In fact, the treatment of severe combined immunodeficient (SCID) mice reconstituted with human T cells and infected with HIV by the injection of MHC-compatible HIV Nef-specific CTL clones leads to the control of HIV replication (KOENIG et al. 1992). In experimentally SIV-infected macaques, the presence of SIV-specific CTL activity correlates with both a reduced efficiency in isolating virus from the blood of these monkeys and with their extended survival (MILLER et al. 1990). In HIV-infected subjects, we have shown a link between the rate of evolution to AIDS-related complex (ARC) and AIDS and the absence of Gag-specific

CTL activity (RIVIÈRE et al. 1991). All these reports support the concept that a candidate AIDS vaccine should also induce a HIV-specific CTL activity.

Acknowledgements. We are grateful to B. Culmann for providing unpublished results and for helpful discussions. We thank L. Chakrabarti and M.L Michel for critical review of the manuscript.

References

Autran B, Letvin NL (1991) HIV epitopes recognized by cytotoxic T-lymphocytes. AIDS 5: S145–S150

Bourgault I, Venet A, Levy JP (1992) Three epitopic peptides of the simian immunodeficiency virus nef protein recognized by macaque cytolytic T lymphocytes. J Virol 66: 750–756

Brinchmann JE, Gaudernack G, Vartdal F (1990) CD8$^+$ T cells inhibit HIV replication in naturally infected CD4$^+$ T cells. J Immunol 144: 2961–2966

Buseyne F, McChesney M, Tanneau F, Montagnier L, Rivière Y (1992) Characterization of the cytotoxic immune responses to HIV-1. In: P Racz, N Letvin, J Gluckman (eds) Cytotoxic T cells in HIV and other retroviral infections. Basel, Karger, pp 81–85

Buseyne F, McChesney M, Porrot F, Kovarik S, Guy B, Rivière Y (1993a) Gag-specific cytotoxic T lymphocytes from human immunodeficiency virus type 1-infected individuals: gag epitopes are clustered in three regions of the p24^{gag} protein. J Virol 67: 694–702

Buseyne F, Blanche S, Schmitt D, Griscelli C, Rivière Y (1993b) Detection of HIV-specific cell mediated cytotoxicity in the peripheral blood from infected children. J Immunol 150: 3569–3581

Chenciner N, Michel F, Dadaglio G, Langlade-Demoyen P, Hoffenbach A, Leroux A, Garcia-Pons F, Rautmann G, Guy B, Guillon JM, Mayaud C, Girard M, Autran B, Kieny MP, Plata F (1989) Multiple subsets of HIV-specific cytotoxic T lymphocytes in humans and in mice. Eur J Immunol 19: 1537–1544

Cheynier R, Langlade-Demoyen P, Marescot M-R, Blanche S, Blondin G, Wain-Hobson S, Griscelli C, Vilmer E, Plata F (1992) Cytotoxic T lymphocytes responses in the pheripheral blood of children born to human immunodeficiency virus-1-infected mothers. Eur J Immunol 22: 2211–2217

Cullen BR (1991) Regulation of HIV-1 gene expression. FASEB J 5: 2361–2368

Culmann B, Gomard E, Kieny MP, Guy B, Dreyfus F, Saimot A-G, Sereni D, Levy JP (1989) An antigenic peptide of the HIV-1 nef protein recognized by cytotoxic T lymphocytes of seropositive individuals in association with different HLA-B molecules. Eur J Immunol 19: 2383–2386

Culmann B, Gomard E, Kieny M-P, Guy B, Dreyfus F, Saimot A-G, Sereni D, Sicard D, Levy JP (1991) Six epitopes reacting with human cytotoxic CD8$^+$ T cells in the central region of the HIV-1 nef protein. J Immunol 146: 1560–1565

Daniel MD, Kirchhoff F, Czajak SC, Sehgal PK, Desrosiers RC (1992) Protective effects of a live attenuated SIV vaccine with a deletion in the *nef* gene. Science 258: 1938–1941

Feinberg MB, Greene WC (1992) Molecular insights into human immunodeficiency virus type 1 pathogenesis. Curr Opin Immunol 4: 466–474

Greene WC (1991) The molecular biology of human immunodeficiency virus type 1 infection. N Engl J Med 324: 308–317

Hadida F, Parrot F, Kieny MP, Sadat-Sowti B, Mayaud C, Debrè P, Autran B (1992a) Carboxyl-terminal and central regions of human immunodeficiency virus-1 *nef* recognized by cytotoxic T lymphocytes from lymphoid organs. J Clin Invest 89: 53–60

Hadida F, Samri A, Hosmalin A, Spohn R, Jung G, Debrè P, Autran B (1992b). Characterization of several promiscuous CTL epitopes in the C-terminal region of Nef. 8th international conference on AIDS, Amsterdam, ThA 1576

Haseltine WA (1991) Molecular biology of the human immunodeficiency virus type 1. FASEB J 5: 2349–2360

Jassoy C, Johnson RP, Navia BA, Worth J, Walker BD (1992) Detection of a vigorous HIV-1 specific cytotoxic T lymphocyte response in cerebrospinal fluid from infected persons with AIDS dementia complex. J Immunol 149: 3113–3119

Kestler III HW, Ringler DJ, Mori K, Panicalli DL, Sehgal PK, Daniel MD, Desrosiers RC (1991) Importance of the *nef* gene for maintenance of high virus loads and for development of AIDS. Cell 65: 651–662

Koenig S, Fuerst TR, Wood LV, Woods RM, Suzich JA, Jones GM, De La Cruz VF, Davey RT, Venkatesan S, Moss B, Biddison WE, Fauci AS (1990) Mapping the fine specificity of a cytolytic T cell response to HIV-1 *nef* protein. J Immunol 145: 127–135

Koenig S, Van Kuyk R, Jones G, Torbett B, Mosier D, Brewah YA, Leath S, Davey VJ, Yannelli J, Rosenberg S, Fauci AS, Lane HC (1992) Adoptive transfer of a nef-specific CTL clone into hu-PBL-SCID mice and an HIV-infected patient. 8th international conference on AIDS, Amsterdam, ThA 1573

Lamhamedi-Cherradi S, Culmann-Penciolelli B, Guy B, Kieny M-P, Dreyfus F, Saimot A-G, Sereni D, Sicard D, Levy JP, Gomard E (1992) Qualitative and quantitative analysis of human cytotoxic T-lymphocyte responses to HIV-1 proteins. AIDS 6: 1249–1258

McChesney M, Tanneau F, Regnault A, Sansonetti P, Montagnier L, Kieny M-P, Rivière Y (1990) Detection of primary cytotoxic T lymphocytes specific for the envelope glycoprotein of HIV-1 by deletion of the env amino-terminal signal sequence. Eur J Immunol 20: 215–220

Michel F, Hoffenbach A, Froussard P, Langlade-Demoyen P, Kaczorek M, Kieny MP, Plata F (1992) HIV-1 *env, nef* and *gag*-specific T-cell immunity in mice: conserved epitopes in *nef* p27 and *gag* p25 proteins. AIDS Res Hum Retrovirus 8: 469–478

Miller MD, Lord CI, Stallard V, Mazzara G, Letvin NL (1990) The gag specific cytotoxic T lymphocytes in rhesus monkeys infected with the simian immunodeficiency virus of macaques. J Immunol 144: 122–128

Mills KHG, Nixon DF, McMichael AJ (1989) T-cell strategies in AIDS vaccines: MHC-restricted T-cell responses to HIV proteins. AIDS 3: S101-S110

Myers G, Berzofsky JA, Korber RF, Smith T, Pavlakis GN (1992) A compilation and analysis of nucleic acid and amino-acid sequences. Los Alamos National Laboratory, Los Alamos, N Mex

Nixon DF, McMichael AJ (1991) Cytotoxic T-cell recognition of HIV proteins and peptides. AIDS 5: 1049–1059

Riviere Y (1990) The cellular immune response to the human immunodeficiency virus. Curr Opin Immunol 2: 424–427

Rivière Y, Tanneau-Salvadori F, Regnault A, Lopez O, Sansonetti P, Guy B, Kieny MP, Fournel JJ, Montagnier L (1989) Human immunodeficiency virus-specific cytotoxic responses of seropositive individuals: distinct types of effector cells mediate killing of targets expressing gag and env proteins. J Virol 63: 2270–2277

Rivière Y, McChesney M, Porrot F, Tanneau F, Buseyne F, Regnault A, Kovarik S, Sansonetti P, Lopez O, Mollereau M, Pialoux G, Marie C, Chamaret S, Kieny MP, Tekaia F, Montagnier L (1991) Relation between HIV disease stages and the activities of HIV specific cytotoxic effector cells. In: Girard M, Valette L (eds) 6è Colloque des Cent Gardes. Mèrieux, Lyon

Robert-Guroff M, Popovic M, Gartner S, Markham P, Gallo RC, Reitz M (1990) Structure and expression of tat, rev and nef specific transcripts of human immunodeficiency virus type 1 in infected lymphocytes and macrophages. J Virol 64: 3391–3398

Robertson MN, Buseyne F, Schwartz O, Rivière Y (1993) Expression of the HIV-1 nef protein by a retroviral vector in an efficient way to present antigen to specific cytotoxic lymphocytes. AIDS Research and human retroviruses 9(12): 1211–1217

Rötzchke O, Falk K, Deres H, Schild H, Norda N, Metzger J, Jüng G, Rammensee HG (1990) Isolation and analysis of naturally processed viral peptides as recognized by cytotoxic T cells. Nature 348: 252–254

Steffy K, Wong-Staal F (1991) Genetic regulation of human immunodeficiency virus. Microbiol Rev 55: 193–205

Tanneau F, McChesney M, Lopez O, Sansonetti P, Montagnier L, and Rivière Y (1990) Primary cytotoxicity against the envelope glycoprotein of human immunodeficiency virus-I: evidence for antibody-dependent cellular cytotoxicity in vivo. J Infect Dis 162: 837–843

Van Bleck GM, and Nathenson SG (1990) Isolation of an endogenously processed immunodominant viral peptide from the class I H-2kb molecule. Nature 348: 213–216

Venet A, Gomard E, Levy J (1993) Human T cell responses to HIV. In: Thomas DB (ed) Viruses and the cellular immune responses. Marcel Dekker. New York, pp 165–200

Walker CM, Moody DJ, Shites DP, and Levy JA (1986) $CD8^+$ lymphocytes can control HIV infection in vitro by suppressing virus replication. Science 234: 1563–1566
Walker CM, Erickson AL, Hsueh FC, Levy JA (1991) Inhibition of human immunodeficiency virus replication in acutely infected $CD4^+$ cells by $CD8^+$ cells involves a noncytotoxic mechanism. J Virol 65: 5921–5927
Zinkernagel RM, Althage A (1977) Antiviral protection by virus-immune cytotoxic T cells: infected target cells are lysed before infectious virus progeny is assembled. J Exp Med 145: 644–651

Cytotoxic T Lymphocytes Specific for Influenza Virus

A. McMichael

1 Introduction

Influenza virus has proved an excellent model for the study of the specificity of cytotoxic T lymphocytes (CTL) and of the processing of viral antigens. Fundamental information about these processes has been gained in both mice and humans. At the same time, a considerable amount has been learned about the role of CTL in influenza viral infections, particularly in mice. The virus has proved to be a malleable tool because it can be readily grown, the full nucleotide structure has been determined (in the 1970s; reviewed in Lamb 1983), there are multiple antigenic variants, many of which have been sequenced, the three-dimensional structures of haemagglutinin and neuraminidase are known (Varghese et al. 1983; Wiley et al. 1981) and, for human studies, it has the advantage that it has infected a very high percentage of the population. All of these features combine to make this a safe and invaluable model for the study of theoretical aspects of T cell function as well as investigating the practical issues of how T cells control virus infections. Studies with influenza virus answered long-standing questions about the interaction between viral antigens and class I molecules of the major histocompatibility complex (MHC). Along the way came information about mutant and variant MHC molecules and the demonstration that peptide

Institute of Molecular Medicine, John Radcliffe Hospital, Oxford, UK

Current Topics in Microbiology and Immunology, Vol. 189

epitopes presented by class I MHC molecules are recognised by CTL. More recently, the virus has been invaluable for the elucidation of the antigen-processing pathways.

2 Influenza as a Model for the Study of Antigen Presentation

2.1 Identification of Peptide Epitopes

Influenza virus is a negative strand RNA virus, whose genome is segmented into eight subunits which can be reassorted in dually infected cells (reviewed in PALESE and YOUNG 1983). The eight RNA segments code for ten proteins, seven of which are present in the virus (LAMB 1983). These are two surface glycoproteins, haemagglutinin and neuraminidase, the viral matrix protein, three RNA polymerases, PB1, PB2 and PA and the nucleoprotein NP. In infected cells three other proteins are expressed, the non-structural proteins NS1 and NS2 and a surface protein encoded by the matrix segment M2. The haemagglutinin and the neuraminidase are highly variable, much of the variation being selected by the humoral immune response (WILEY et al. 1981). Human infections are caused by type A and type B viruses, which, although similar, are structurally quite distantly related. Antigenic differences in the surface glycoproteins of type A virus are the basis of the division of the virus into the major subtypes H1N1, H2N2 and H3N2. When these subtypes first appeared, they caused pandemic influenza because of poor or absent pre-existing humoral immunity in humans; H1N1 virus is thought to have caused the 1919 pandemic, H2N2 caused the Asian influenza of 1957 and H3N2 the Hong Kong influenza of 1968. In subsequent years there was antigenic drift in both haemaglutinin and neuraminidase of each subtype, selected by the humoral immune response in human populations. Such strains are normally identified by their year and place of isolation, e.g. A/Puerto Rico/1934 (A/PR/8/34, an H1N1 virus). The virus can infect most mammalian cell types, but the infection is productive in only a few. This appears to relate to the cleavage of the haemagglutinin into subunits HA1 and HA2, which is essential for virus infectivity and function (GARTEN et al. 1989).

CTL specific for influenza virus were first demonstrated in mice in 1977 (ZWEERINK et al. 1977) and shortly after in humans (MCMICHAEL et al. 1977). Similar strategies were used, lymphocytes were removed from previously infected individuals, restimulated with autologous virus infected cells for 7–14 days and then tested for lytic activity on virus-infected target cells. MHC class I restriction, the hallmark of CTL-mediated lysis of target cells (ZINKERNAGEL and DOHERTY 1979), was demonstrated in both mice and humans. An early finding was that CTL did not discriminate between major virus subtypes; for instance, H3N2-specific CTL-lysed H1N1 infected cells equally well even when human donors were used who had never been infected with the latter virus (McMichael and Askonas 1978). At first it was thought that CTL recognized shared epitopes in the surface glycoproteins, such as haemagglutinin, even though these were in some cases only 30% homologous; later experiments showed that the CTL were largely recognising conserved internal virus proteins (see below). Influenza-specific CTL proved to be exquisitely sensitive to small changes in MHC structure. CTL failed to recognise viral antigen presented by serologically defined HLA A2 molecules (McMichael 1978), which were then shown to differ in amino acid sequence (Biddison et al. 1980; Krangel et al. 1982); subsequently, similar findings were made for other HLA antigens such as HLA B27 (Breuning et al. 1982; Parham et al. 1988). In each case, the variant HLA class I molecules differ in only one to four amino acids. These variants are therefore very like the mutants of H-2 K^b, both structurally and functionally (Nathenson et al. 1986).

In 1981 the first CTL clones specific for influenza virus were derived (Lin and Askonas 1981). Like the polyclonal cultures, most of these were fully cross-reactive when tested with different influenza virus strains. Of the strain-specific clones, some recognised HA (Lukacher et al. 1984), but two clones were identified that distinguished between 1934 and 1968 viruses (Bennink et al. 1982; Townsend and Skehel 1982). Using reassorted viruses, in which RNA segments had been mixed, both Bennink et al. and Townsend et al. mapped the specificity of their CTL clones. The former mapped to PB2 polymerase and the latter to NP, both of which differed in a few amino acids between the 1934 and 1968 virus strains. These results were followed up by testing CTL against target cells transfected with cDNA corresponding to single virus products. Braciale et al. (1984) had demonstrated HA-specific CTL, but Townsend et al. demonstrated that his clone was NP specific and, in the same study, that the majority of polyclonal CTL in H-2^b mice were specific for NP (Townsend et al. 1984). Subsequently, the majority of murine and human CTL were shown to react with NP, PB2, HA and M (Bennink et al. 1986; Yewdell et al. 1985; Gotch et al. 1987a); it is of interest that the anti-M response, which is prominant in humans (Gotch et al. 1987), is barely found in mice, although the murine cell line P815 transfected with HLA A2 can readily present virus to human matrix protein-specific CTL (Rowland-Jones and McMichael, unpublished).

These findings raised the question of how an internal protein was presented to CTL when it could not be detected intact on the surface of infected cells. Arguing that the internal virus proteins might be processed in a similar way to protein antigens presented by class II MHC, Townsend et al. transfected fragments of NP into target cells and demonstrated that fragments were recognized by CTL, but that different fragments appeared to be presented by different MHC class I antigens (Townsend et al. 1985).

GOODING and O'CONNELL (1983) had previously carried out a similar experiment with CTL specific for the T antigen of SV40. It was later demonstrated that cells transfected with cDNA for haemaglutinin with the signal sequence deleted lysed those targets very efficiently, even though HA was absent from the cell surface (TOWNSEND et al. 1986a). From all of these results it was argued that CTL might recognize cytoplasmic proteins degraded to peptide fragments. Therefore, peptides were added to uninfected cells and it was found that CTL lysed the targets (TOWNSEND et al. 1986b). Furthermore, in the initial experiments it was clear that H-2Db and HLA B37 presented different peptides from NP. These results implied that the class I MHC molecules bound antigenic peptides in a specific fashion and presented them to CTL.

Shortly after these findings were made, the crystal structure of HLA A2 was solved and unidentified electron density, thought (and now known) to be a mixture of peptides, was present in a cleft on the membrane-distal surface of the molecule (BJORKMAN et al. 1987a, b). Most of the polymorphism in the MHC class I molecules was in residues whose side chains contributed to this groove, and as further class I molecules, HLA A68, B27 and H-2K^b, have been solved (FREMONT et al. 1992; GARRETT et al. 1989; MADDEN et al. 1992; ZHANG et al. 1992), it has become clear that there are fine differences in the grooves of different class I molecules, explaining why they bind different peptide epitopes.

Several epitopes have been identified in the various proteins of influenza virus that are presented by different MHC class I molecules; these are summarised in Table 1. It is clear that different class I MHC molecules select different peptide epitopes. Peptides have been eluted from different class I MHC molecules (FALK et al. 1991b; HUNT et al. 1992; JARDETZKY et al. 1991), and it is clear that peptides binding to each are short and have different sequence motifs. Thus, peptides binding to HLA A2 have leucine or isoleucine at position 2 and leucine, isoleucine or valine at position 9 (HUNT et al. 1992) and often an aromatic residue at position 3 (PARKER et al. 1992); peptides binding to HLA B27 have arginine at position 2, often an aromatic residue at position 3 and usually an arginine or lysine at position 9 (JARDETZKY et al. 1991). Peptides are most commonly nine amino acids in length, but the range is from eight to 12. Experiments on presentation of influenza virus by site-directed mutant MHC molecules orientated peptides with the amino terminus on the left and the carboxyl terminus on the right (LATRON et al. 1991). For instance, the introduction of an arginine at position 116 in the floor of the groove in HLA A2 impaired presentation of the influenza matrix peptide 58–66, but normal titration could be restored by changing position 9 to arginine. More recently, exchange of six amino acids which form the B pocket in the groove from HLA B27 sequence to HLA A2 sequence changed the specificity of the second anchor residue in the peptide from arginine to leucine (COLBERT et al. 1993). Fine structural analysis has revealed that the amino terminus binds into the A pocket at the left-hand end of the groove, where there are four conserved tyrosines (MADDEN et al. 1992).

Table 1. Cytotoxic T lymphocyte epitopes in influenza A virus

Virus Protein	First residue	Amino acid sequence	MHC restriction	References
Nucleoprotein (NP)	50-	SDYEGRLI	H-2K^k	Bastin et al. 1987; Cossins et al. 1993
	91-	KTGGPIYKR	HLA Aw68	Cerundolo et al. 1991; Guo et al. 1992
	147-	TYQRTRALV	H-2K^d	Bodmer et al. 1988; Falk et al. 1991a
	342-	FEDLRVLSFI	HLA B37	McMichael et al. 1986
	366-	ASNENMETL	H-2D^b	Falk et al. 1991a; Townsend et al. 1986b
	380-	ELRSRYWAI	HLA 38	Sutton et al. 1993
	383-	SRYWAIRTR	HLA B2705	Huet et al. 1990
Matrix protein (M1)	58-	GILGFVFTL	HLA A2	Gammon et al. 1992; Gotch et al. 1987b; Morrison et al. 1992
Haemagglutinin (HA)	204-	LYQNVGTYV	H-2 K^d	Hahn et al. 1992
	210-	TYVSVGTSTL	H-2 K^d	Hahn et al. 1992
	259-	FEANGNLI	H-2 K^k	Gould et al. 1991
(HA2)	10-	IEGGWTGMI	H-2 K^k	Gould et al. 1991
		IYSTVASSL	H-2 K^d	Milligan et al. 1990
Non-structural protein-1	152-	EEGAIVGEI	H-2 K^k	Cossins et al. 1993

These form a hydrogen bond network with the amino terminus; disruption of this network by changing one of these tyrosines to phenylalanine greatly peptide presentation (LATRON et al. 1992). Similarly at the C terminus, hydrogen bonds are formed in the F pocket with Thr-84 and Tyr-143 (MADDEN et al. 1992). The most detail is known about how the HLA A68 presents peptide NP101–109. HLA A68 has been crystallised with this peptide in the groove and the three-dimensional structure determined (GUO et al. 1992; SILVER et al. 1992). Again, the amino terminus fits in the A pocket and the carboxyl terminus in the F pocket, the interactions being very similar to those described for HLA B27. Because of the conservation of the A and F pockets in classical class I molecules, it is likely that they all bind peptides in the same way.

Experiments on influenza virus have also revealed details of antigen processing. As more epitopes are determined, there appears to be some clustering of epitopes in influenza NP (Table 1). This may indicate sites of preferential cleavage, perhaps related to the three-dimensional structure of NP. Similar clustering has been found in HIV gag and HIV nef (CULMANN et al. 1991), two other proteins in which multiple epitopes are known (McMichael et al., unpublished). Clustered epitopes in restricted regions mean that many overlap, with possible consequences in terms of competition for processing pathways or binding to class I molecules.

2.2 Processing of Influenza Virus Proteins

TOWNSEND and colleagues argued that peptides generated in the cytoplasm need to be transported into the endoplasmic reticulum (ER), where they would meet the MHC molecules (TOWNSEND and BODMER 1988). Because epitopes were found in many different places in given proteins, they could not have common signal sequences. This implied a role for transporters to take peptides across the ER membrane. A mutant cell line, RMA-S, was found where class I MHC was not expressed on the cell surface but unfolded heavy chains were present in the ER (TOWNSEND et al. 1989b). Addition of influenza NP epitope peptides to these cells stabilised expression of class I molecules on the surface, probably by binding to unstable empty molecules (TOWNSEND et al. 1989a). Thus, these cells could present added peptide, but when infected with influenza virus, CTL did not react. Added peptide is thought to bind to empty class I molecules on the surface of cells because they can still bind and present in the presence of brefeldin A, which inhibits passage of proteins through the Golgi network (NUCHTERN et al. 1989; YEWDELL and BENNINK 1989). In the RMA-s cell line, empty class I molecules come out to the surface in an unstable form that can be stabilised by cooling, addition of anti-MHC antibody or binding a (influenza NP) peptide epitope (LJUNGGREN et al. 1990). All of these data implied that there was a fault in the antigen-processing pathway in RMA-S. A human mutant cell line, 721. 174, and its derivative T2, showed a similar phenotype (CERUNDOLO et al. 1990). This fault could be corrected by transfecting with a minigene for a peptide, influenza matrix 57–68, encompassing the epitope presented by HLA A2 (58–66), provided it had a signal sequence to take it into the ER (ANDERSON et al. 1991); this again supported the assumption that the fault was in the delivery of peptide into the ER. The T2/721.174 cell lines have a large deletion in the class II region of the MHC and this region was therefore targeted in the search for possible transporter genes. Two, *TAP-1* and *TAP-2*, which encode the two chains of an adenosine triphosphate (ATP)-binding cassette transporter protein (SPIES et al. 1992), were found in this region in HLA and the equivalent region of H-2 (DEVERSON et al. 1990; MONACO et al. 1990; SPIES et al. 1990; TROWSDALE et al. 1990), and transfection of these genes into mutant cell lines corrected the defect (SPIES et al. 1992). It is clear, therefore, that the transporters are important in the processing of cytoplasmic antigens. However, it is still not absolutely certain that they actually transport peptides; LEVY et al. (1991) using a microsome preparation found that added peptides could enter the microsomes without the need for ATP. Despite this finding, it is widely held that the *TAP* gene products actually transport peptides.

It is very clear that the antigen-processing pathway for peptides presented by class I and II MHC is very different. MORRISON et al. (1986) first showed differential susceptibility of the processing of HA epitopes to class I- and II-restricted T cell clones to chloroquine and fixation of antigen-presenting cells. Later, it was shown that T2 cells retained the ability to present to class II-restricted T cells, while presentation through class I was lost (ZWEERINK et al. 1993).

Once in the ER, peptides bind to MHC class I molecules. TOWNSEND has demonstrated that they stabilise newly folded class I molecules (ELLIOTT et al. 1991; TOWNSEND et al. 1990). Binding affinities are greatest for optimum length peptides (ELLIOTT et al. 1991), which can bind to free heavy chain in the absence of β_2-microglobulin. It is not clear, however, where the optimum length peptide is generated. A likely candidate for mediating the proteolytic cleavage is the multi-catalytic protease complex (GLYNNE et al. 1991; KELLY et al. 1991; MARTINEZ and MONACO 1991; ORTIZ-NAVARETTE et al. 1991). Two components of the 20–30 that make up the particle are encoded within the MHC, although in experimental systems, the T2 cell line with the transporters transfected back, but still deficient in these two genes, present antigen perfectly well (MOMBURG et al. 1992). If this complex generates peptides which are transported into the ER, it is likely they are cleaved to the optimum length within the ER. The known specificity of proteases is to clip on the C-terminal side of hydrophobic or charged residues (DJABALLAH and RIVETT 1992). In an analysis of more than 50 peptide epitopes that have been carefully defined, we have noted that the C terminus is highly preserved ELLIOTT et al. (1983). Isoleucine and leucine are the commonest residues, followed by arginine and lysine and, in some cases, tyrosine and phenylalanine.

Occasionally, phenylalanine and valine have been found; the majority of amino acids have never been observed. The three types of C terminus correlate with the pocket close to the F pocket in the peptide-binding groove, with position 116 playing a key role. If this is tyrosine (as in HLA A2), the C terminus tends to be isoleucine, leucine or valine (in descending order of frequency). If residue 116 is aspartic acid (as in HLA B27), arginine or lysine are preferred, although isoleucine is sometimes found. If 116 is serine, tyrosine is preferred, possibly forming a hydrogen bond. These requirements are not absolute and there are exceptions, but these are still within this limited range of amino acids. Therefore, it has been suggested that the C terminus is largely determined by the proteases (ROTZSCHKE and FALK 1991; ELLIOTT et al. 1993). Peptides would then be delivered into the ER, by the transporters, with hydrophobic or positively charged termini. Further modification of the peptides would occur at the N terminus, either before or after peptide has bound. It has been suggested that the MHC-encoded protease components divert the protease activity away from cleaving after acidic residues; acidic C termini have not yet been found in epitope peptides (Gaczynska et al. 1993). There is some evidence that the amino acid sequences flanking the epitope may affect processing (CERUNDOLO et al. 1991; EISENLOHR et al. 1992), but other evidence implies that the position of the epitope in the protein has little effect (HAHN et al. 1992). Overall, there is still much to learn about the generation of the epitope peptides and it should be emphasised that there is still no firm evidence that the proteasome is responsible for proteolytic cleavage of viral proteins in the cytoplasm of infected cells.

2.3 T Cell Receptors Used by Anti-Influenza CTL

T cell receptors used by influenza-specific CTL have been analysed by determining the receptors of cloned CTL (BOWNESS et al. 1993; MOSS et al. 1991). MOSS et al. found that for CTL recognising influenza matrix protein and presented by HLA A2, there was a very large predominance of T cell clones with $V_\alpha10$ and $V_\beta17$. More recent results have confirmed that $V_\beta17$ occurs in about 70% of all T cell clones that recognise this particular peptide epitope (Lehner and Borysiewicz, personal communication). Very similar receptors, differing in only one or two conservative amino acids, have been found in receptors generated from different individuals. This is a remarkable result considering the enormous diversity of the total T cell receptor repertoire, with over 10^{14} possibilities (DAVIS and BJORKMAN 1989). Similar results were obtained by Bowness et al. for HLA B27-restricted CTL recognising an NP peptide. Here, $V_\beta7$ and 8 and $V_\alpha12$, 14 and 22 were preferred with, again, very similar receptors in clones derived from different individuals. Furthermore, in both the HLA A2- and B27-restricted responses there was marked conservation of J_α and conservation of the junctional region in the

β-chain, normally the most variable part of the receptor because of the contributions of V_β, D_β, J_β plus random N-diversity. These results imply that T cell receptor usage of HLA class I virus-specific CTL is highly restricted, although even in very similar receptors, fine specificity differences can be identified (Bowness et al. and Haurum et al., unpublished). These results have at least one important implication, namely that in HLA-associated disease, different individuals may use very similar receptors and have similar fine specificities and cross-reactions. In the context of influenza, the findings also suggest at new explanation for the lethal combination of staphylococcal and influenza virus infecion (ROBERTSON et al. 1958), namely that super-antigens may subvert the reacting, clonally restricted, T cells.

3 Influenza-Specific CTL in Recovery from Infection

YAP et al. (YAP and ADA 1978b) were the first to transfer CTL-enriched cultures to mice with influenza pneumonitis. CD8-positive cells were found to reduce lung virus titres, while CD8-depleted cells did not. These experiments were repeated and extended later with influenza-specific CTL clones with similar results (LIN and ASKONAS 1981; LUKACHER et al. 1984). Both NP-specific and HA-specific clones mediate the effect. Nude mice are vulnerable to influenza infection; they have a higher mortality and survivors secrete virus chronically (WELLS et al. 1981a). Nude mice recovered when CTL were transferred. In these experiments passive antibody failed to clear virus (WELLS et al. 1981b). All of these experiments point to a role for CTL in the recovery phase of influenza virus infection, a conclusion supported by histological inspection of recovering lungs (MACKENZIE et al. 1989). It is obvious that CTL cannot neutralise free virus, but by killing infected cells before the generation of new virus particles CTL could terminate virus replication. It is also possible that CTL release γ-interferon with direct effects on virus replication in infected cells as well as activating natural killer cells and causing up-regulation of class I MHC antigens to enhance presentation (MORRIS et al. 1982). Recently, there has been a discrepant result; Tite et al. immunised mice with a recombinant NP preparation and found good protection from infection with live virus in the absence of a CTL response, but in the presence of a $CD4^+$ T cell response (TITE et al. 1990a). Now that it is known that CD4 T cells can be divided into Th1 and Th2 subsets according to the pattern of cytokines secreted, it will be important to reassess the roles of these subsets protection. It is likely that in real life several types of immune response will combine to give protective immunity.

YAP et al. (YAP and ADA 1978a) measured the CTL response in mouse spleens early in influenza virus infection. They found that CTL were measurable before the primary IgM antibody response. Thus, CTL could be largely responsible for the initial clearance of virus early in an infection. In some infections, this phase is associated with briefly enhanced symptoms, for instance a fever and rash. It should also be pointed out that although in other infections CTL may contribute to immunopathology (BUCHMEIER et al. 1980; CANNON et al. 1988; KRETH et al. 1982; PLATA et al. 1987; SETHI et al. 1988), this does not appear to be the case in influenza infection.

In humans, CTL activity was measured in volunteers who were challenged intranasally with influenza virus (MCMICHAEL et al. 1983b). There was a correlation between the levels of CTL activity and clearance of intranasal virus, measured 3 and 4 days later. Anti-haemagglutinin antibody also protected, but there were individuals who appeared to be protected by CTL in the absence of any neutralising antibody. Similarly, patients with hypogammaglobulinaemia appear to have handled pandemic influenza normally, despite any protection from administered γ-globulin (MACCALLUM 1971).

How does this fit in to the epidemiology of influenza virus infection in humans, particularly the observation that humans are repeatedly infected with the virus throughout life? Neutralising antibody is primarily directed against the HA and selects HA variants in the population (WEBSTER et al. 1982; WILEY et al. 1981). Thus, protection by antibody becomes progressively less efficient as the years go by. In contrast, human CTL recognise the less variable internal virus proteins particularly NP and M1 (GOTCH et al. 1987a). In vitro, CTL cross-react between viruses of different subtypes. Thus, a good CTL memory response might protect against most strains of virus. Although CTL cannot protect against infection, they should clear virus more rapidly resulting in less severe disease, as in the volunteer studies (MCMICHAEL et al. 1983b). However, clinical experience is that humans are susceptible to several moderate to severe attacks of influenza in a lifetime. The probable explanation is that in the absence of persisting viral antigen, CTL memory declines after acute infection. When CTL responses in naturally infected humans were observed over time such a decline was observed, giving an approximate half-life of 4 years for influenza-specific CTL (MCMICHAEL et al. 1983a). Recently, more evidence has been obtained in mice to suggest that CTL memory requires persisting viral antigen (OEHEN et al. 1992).

The real test of CTL immunity in humans comes when a new pandemic virus strain appears, as occurred in 1919, 1957 and 1968. A very high proportion of humans are susceptible at these times because of a complete absence of protective antibody (except in the elderly) and the mortality is high. Yet, at the same time many individuals have subclinical infections (FOY et al. 1976); perhaps many of these individuals are protected by cross-reactive CTL and clear virus rapidly (ASKONAS et al. 1982). This hypothesis can be tested when a new pandemic strain emerges.

4 Influenza Vaccine Design

This knowledge of cellular immunity to influenza virus infection has implications for vaccine design. Most of the influenza vaccines in use are killed virus preparations or subunit vaccines. These cannot be expected to deliver cytoplasmic antigens efficiently to the class I antigen-processing system. Studies on vaccines in mice and humans showed that they are poor at stimulating CTL responses (McMichael et al. 1981; Webster and Askonas 1980). Influenza vaccines are not entirely satisfactory. At least one study has shown that over several years, annual vaccination is slightly less effective than no vaccination followed by natural infection (Hoskins et al. 1979).

Knowledge of antigen processing and peptide epitope presentation to CTL suggests new ways of vaccinating to maximise the CTL response. Recombinant virus vaccines such as vaccinia are effective in inducing CTL responses in animals and humans (Bennink et al. 1984; Gotch et al. 1991); however, their use for influenza vaccination is unlikely, except possibly in the face of a very severe pandemic. Recombinant bacteria such as oral salmonella have been shown to induce CTL responses to influenza NP, but rather inefficiently (Gao et al. 1991, 1992). Recombinant proteins and killed virus preparations are not efficient in generating CTL responses (Tite et al. 1990b; Webster and Askonas 1980), probably because the injected materials fail to find their way into the class I MHC-processing pathway. An alternative approach would be to use peptides to induce a pure CTL response, although multiple peptides would have to be used to cover the common HLA types and hence the majority of the population. Immunisation with peptides alone can induce CTL responses, but not at a very high level (Gao et al. 1991). Modification of the injected peptide by addition of a lipid tail (Deres et al. 1989) or mixing with immune-stimulating complexes (ISCOM; Berzofsky 1991; Takahashi et al. 1990) appears to make the immunisation more efficient. An interesting new approach has been the injection of cDNA for NP intramuscularly, which primes for CTL responses in mice (Ulmer et al. 1993). These issues are being addressed in a number of virus infections where vaccines are urgently needed such as HIV, hepatitis C and malaria; much of the knowledge gained on priming CTL responses with recombinant material will be common to all of these infections.

5 Conclusions

Influenza virus has proved to be an invaluable tool for unravelling the mechanisms by which internal antigens are processed and presented to CTL. Studies on infection in vivo have led to a better understanding of the role of

CTL in acute virus infections in general, as well as influenza. These insights should lead to better vaccines in the near future.

References

Anderson K, Cresswell P, Gammon M, Hermes J, Williamson A, Zweerink H (1991) Endogenously synthesized peptide with an endoplasmic reticulum signal sequence sensitizes antigen processing mutant cells to class I-restricted cell-mediated lysis. J Exp Med 174 (2): 489–492

Askonas BA, McMichael AJ, Webster, RG (1982) The immune response to influenza virus and the problem of protection against infection. In: Beare AS (eds) Basic and applied influenza research CRC Press, Boca Raton, pp 157–188

Bastin J, Rothbard J, Davey J, Jones I, Townsend, A (1987) Use of synthetic peptides of influenza nucleoprotein to define epitopes recognized by class I restricted cytotoxic T lymphocytes. J Exp Med 165: 1508

Bennink JR, Yewdell JW, Gerhard W (1982) A viral polymerase involved in recognition of influenza virus-infected cells by a cytotoxic T cell clone. Nature 296: 75–76

Bennink JR, Yewdell JW, Smith GL, Moller C, Moss B (1984) Recombinant vaccinia virus primes and stimulates influenza haemagglutinin-specific cytotoxic T cells. Nature 311 (5986): 578–579

Bennink JR, Yewdell JW, Smith GL, Moss B (1986) Anti-influenza cytotoxic T lymphocytes recognise the three viral polymerases and a non-structural protein: reponsiveness to individual viral antigens is MHC controlled. J Virol 61: 1098–1102

Berzofsky JA (1991) Mechanisms of T cell recognition with application to vaccine design. Mol Immunol 28(3): 217–223

Biddison WE, Ward FE, Shearer GM, Shaw S (1980) The self determinants recognised by human virus immune T cells can be distinguished from the serologically defined HLA antigens. J Immunol 124: 548–552

Bjorkman P, Saper M, Samraoui B, Bennett W, Strominger J, Wiley D (1987a) Structure of human class I histocompatibility antigen, HLA-A2 Nature 329: 506–511

Bjorkman PJ, Saper MA, Samraoui B, Bennett WS, Strominger JL, Wiley DC (1987b) The foreign antigen binding site and T cell recognition regions of class I histocompatibility antigens. Nature 329: 512–518

Bodmer HC, Pemberton RM, Rothbard JB, Askonas BA (1988) Enhanced recognition of a modified peptide antigen by cytotoxic T lymphocytes specific for influenza nucleoprotein. Cell 52: 253–258

Bowness PA, Moss PAH, Rowland-Jones SL, Bell JI, McMichael AJ (1993) Conservation of T cell receptor usage by HLA B27-restricted influenza specific cytotoxic T lymphocytes suggests a general pattern for antigen-specific MHC class-I restricted responses. Eur J Immunol (in press)

Braciale TJ, Braciale VL, Henkel TJ, Sambrook J, Gething M-J (1984) Cytotoxic T lymphocyte recognition of the influenza haemagglutinin gene product expressed by DNA mediated gene transfer. J Exp Med 159: 341–349

Breuning MH, Lucas CJ, Breur BJ, Engelsma MY, DeLange GG, Dekker AJ, Biddison WE, Ivanyi P (1982) Subtypes of HLA B27 detected by cytotoxic T lymphocytes and their role in self recognition. Hum Immunol 5: 259–268

Buchmeier MJ, Welsh RM, Dutko FJ, Oldstone MBA (1980) The virology and immunology of lymphocytic choriomeningitis virus infection. Adv Immunol 30: 275–331

Cannon MJ, Openshaw PJM, Askonas B (1988) Cytotoxic cells clear virus but augment lung pathology in mice infected with respiratory syncytial virus. J Exp Med 168: 1163–1168

Cerundolo V, Alexander J, Anderson K, Lamb C, Cresswell P, McMichael A, Gotch F, Townsend A (1990) Presentation of viral antigen controlled by a gene in the major histocompatibility complex. Nature 345 (6274): 449–452

Cerundolo V, Tse AGD, Salter RD, Parham P, Townsend A (1991) CD8 independence and

specificity of cytotoxic T lymphocytes restricted by HLA Aw68.1. Proc R Soc Lond [B] 244: 169–177
Colbert R, Rowland-Jones S, McMichael AJ, Frelinger J (1993) Proc Natl Acad Sci USA (in press)
Cossins J, Gould KG, Smith M, Driscoll P, Brownlee, GG (1993) Precise prediction of a Kk-restricted cytotoxic T cell epitope in the NS1 protein of influenza virus using an MHC allele-specific motif. Virology 193 (1): 289–295
Culmann B, Gomard E, Kieny M-P, Guy B, Dreyfus F, Saimot A-G, Sereni D, Sicar D, Levy J-P (1991) Six epitopes reacting with human cytotoxic $CD8^+$ T cells in the central region of the HIV nef protein. J Immunol 146: 1560–1565
Davis MM, Bjorkman PJ (1989) T-cell receptor genes and T-cell recognition. Nature 334: 395–402
Deres K, Schild H, Weissmuller K-H, Jung G, Rammensee H-G (1989) In vivo priming of virus specific cytotoxic T lymphocytes with synthetic lipopeptide vaccine. Nature 342: 561–564
Deverson EV, Gow JR, Coadwell WJ, Monaco JJ, Butcher GW, Howard JC (1990) MHC class II region encoding proteins related to the multidrug resistance family of transmembrane transporters. Nature 348: 738–741
Djaballah H, Rivett AJ (1992) Peptidylglutamyl-peptide hydrolase activity of the multicatalytic proteinase complex: evidence for a new high-affinity site, analysis of cooperative kinetics, and the effect of manganese ions. Biochemistry 31 (16): 4133–4141
Eisenlohr LC, Yewdell JW, Bennink JR (1992) Flanking sequences influence the presentation of an endogenously synthesized peptide to cytotoxic T lymphocytes. J Exp Med 175: 481–487
Elliott T, Cerundolo V, Elvin J, Townsend A (1991) Peptide induced conformational change of a class I heavy chain. Nature 351: 402
Elliott T, Smith M, Driscoll P, McMichael A (1993) Peptide selection by class I molecules of the major histocompatibility complex. Current Biol (in press)
Falk K, Rötzchke O, Deres K, Metzger J, Jung G, Rammensee H-G (1991a) Identification of naturally processed viral nonapeptides allows their quantification in infected cells and suggests an allele-specific T cell epitope forecast. J Exp Med 174: 425–434
Falk K, Rotzschke O, Stevanovic S, Jung G, Rammensee, H-G (1991b) Allele specific motifs revealed by sequencing of self peptides eluted from MHC molecules. Nature 351: 290–296
Foy HM, Cooney MK, Allan I (1976) Longitudinal studies of types A and B influenza among Seattle schoolchildren and their families, 1968–1974. J Infect Dis 134: 362–369
Fremont DH, Matsumura M, Stura EA, Peterson PA, Wilson IA (1992) Crystal structures of two viral peptides in complex with murine MHC class I H-2Kb. Science 257(5072): 919–927
Gaczynska M, Rock KL, Goldberg AL (1993) γ-interferon and expression of MHC genes regulate peptide hydrolysis by proteasomes. Nature 365: 264–267
Gammon MC, Bednarek MA, Biddison WE, Bondy SS, Hermes JD, Mark GE, Williamson AR, Zweerink HJ (1992) Endogenous loading of HLA-A2 molecules with an analog of the influenza virus matrix protein-derived peptide and its inhibition by an exogenous peptide antagonist. J Immunol 148(1): 7–12
Gao XM, Zheng B, Liew FY, Brett S, Tite J (1991) Priming of influenza virus-specific cytotoxic T lymphocytes vivo by short synthetic peptides. J Immunol 147(10): 3268–3273
Gao X-M, Tite JP, Lipscombe M, Rowland-Jones S, Ferguson DJP, McMichael AJ (1992) Recombinant Salmonella typhimurium invading nonphagocytic cells are resistant to recognition by antigen specific cytotoxic T lymphocytes. Infect Immun 60: 3780–3789
Garrett TPJ, Saper MA, Bjorkman PJ, Strominger JL, Wiley DC (1989) Specificity pockets for the side chains of peptide antigens in HLA-Aw68. Nature 342: 692
Garten W, Stieneke A, Shaw E, Wikstrom P, Klenk H-D (1989) Inhibition of proteolytic activation of influenza virus hemagglutinin by specific peptidyl chloralkyl ketones. Virology 172: 25–31
Glynne R, Powis SH, Beck S, Kelly A, Kerr L-A, Trowsdale J (1991) A proteasome-related gene between the two ABC transporter loci in the class II region of the human MHC. Nature 353: 357–359
Gooding LR, O'Connell KA (1983) Recognition by cytotoxic T lymphocytes of cells expressing fragments of the SV40 tumour antigen. J Immunol 131: 2580–2586
Gotch FM, McMichael AJ, Smith GL, Moss B (1987a) Identification of the virus molecules recognised by influenza specific cytotoxic T lymphocytes. J Exp Med 165: 408–416
Gotch F, Rothbard J, Howland K, Townsend A, McMichael A (1987b) Cytotoxic T lymphocytes recognise a fragment of influenza virus matrix protein in association with HLA-A2. Nature 326: 881–882

Gotch FM, Hovell R, Delchambre M, Silvera P, McMichael AJ (1991) Cytotoxic T lymphocyte response simian immunodeficiency virus by cynomolgus macaque monkeys immunized with recombinant vaccinia virus. AIDS 5: 317–320
Gould KG, Scotney H, Brownlee GG (1991) Characterization of two distinct major histocompatibility complex class I Kk-restricted T-cell epitopes within the influenza A/PR/8/34 virus hemagglutinin. J Virol 65(10): 5401–5409
Gould KJ, Scotney H, Townsend ARM, Bastin J, Brownlee GG (1987) Mouse H-2k-restricted cytotoxic T cells recognize antigenic determinants in both the HA1 and HA2 subunits of the influenza A/PR/34 hemagglutinin. J Exp Med 166: 693
Guo H-C, Jardetzky TS, Garrett TP, Lane WS, Strominger JL, Wiley DC (1992) Different length peptides dind to HLA Aw68 similarly at their ends but bulge out in the middle. Nature 360: 364–366
Hahn YS, Hahn CS, Braciale VL, Braciale TJ, Rice CM (1992) $CD8^+$ T cell recognition of an endogenously processed epitope is regulated primarily by residues within the epitope. J Exp Med 176(5): 1335–1341
Hoskins TW, Davies JR, Smith AJ, Miller CL, Allchin A (1979) Assessment of inactivated influenza A vaccine after three outbreaks of influenza A at Christ's Hospital. Lancet i: 33–35
Huet S, Nixon DF, Rothbard J, Townsend ARM, Ellis SA, McMichael AJ (1990) Structural homologies between two HLA B27 restricted peptides suggest residues important for interaction with HLA B27. Int Immunol 2: 311–316
Hunt DF, Henderson RA, Shabanowitz J, Sakaguchi K, Michel H, Sevilir N, Cox AL, Appella E, Engelhard VH (1992) Characterization of peptides bound to the class I MHC molecule HLA A2.1 by mass spectometry. Science 255: 1261–1263
Jardetzky TS, Lane WS, Robinson RA, Madden DR, Wiley DC (1991) Identification of self peptides bound to purified HLA-B27. Nature 353: 326–329
Kelly A, Powis SH, Glynne R, Radley E, Beck S, Trowsdale J (1991) Second proteasome-related gene in the human MHC class II region. Nature 353: 667–668
Krangel MS, Taketani S, Biddison WE, Strong DM, Strominger JL (1982) Comparative structural analysis of HLA-A2 antigens distinguishable by cytoxic T lymphocytes: variants M7 and DR1. Biochemistry 21: 6313–6321
Kreth HW, Kress L, Kress HG, Ott HF, Eckert G (1982) Demonstration of primary cytotoxic T cells in venous blood and cerebrospinal fluid in children with mumps meningitis. J Immunol 128: 2411–2415
Lamb RA (1983) The influenza virus RNA segments and their encoded proteins. In: Palese P, Kingsbury D (eds) Genetics of influenza virus. Springer, Vienna New York, pp 21–69
Latron F, Moots R, Rothbard J, Garrett T, Strominger JL, McMichael AJ (1991) Positioning of a peptide in the cleft of HLA A2 by complementing amino acid changes. Proc Natl Acad Sci USA 88: 11325–11329
Latron F, Pazmany L, Morrison JRM, Saper M, McMichael AJ, Strominger JL (1992) A critical role for conserved residues in the cleft of HLA-A2 in presentation of a nonapeptide to T cells. Science 257: 964–967
Levy F, Gabathuler R, Larsson R, Kvist S (1991) ATP is required for in vitro assembly of MHC class I antigens but not for transfer of paptides across the ER membrane. Cell (in press)
Lin Y, Askonas BA (1981) Biological properties of an influenza A virus specific killer T cell clone. J Exp Med 154: 225–234
Ljunggren HG, Stam NJ, Ohlen C, Neefjes JJ, Hoglund P, Heemels MT, Bastin J, Schumacher TN, Townsend A, Karre K et al. (1990) Empty MHC class I molecules come out in the cold. Nature 346(6283): 476–480
Lukacher AE, Braciale VL, Braciale TJ (1984) In vivo effector function of influenza virus specific cytotoxic T lymphocyte clones is highly specific. J Exp Med 160: 814–826
MacCallum FO (1971) Hypogammaglobulinaemia in the United Kingdom. 7. The role of humoral antibodies in protection against and recovery from bacterial and viral infections in hypogammaglobulinaemia. Med Res Engl 310: 72–85
Mackenzie CD, Taylor PM, Askonas BA (1989) Rapid recovery of lung histology correlates with clearance of influenza virus by specific $CD8^+$ cytotoxic T cells. Immunology 67: 375–381
Madden DR, Gorga JC, Strominger JL, Wiley DC (1992) The three dimensional structure of HLA-B27 at 2.1 Å resolution suggests a general mechanism for tight peptide binding to MHC. Cell 70: 1035–1048
Martinez CK, Monaco JJ (1991) Homology of the proteasome subunits to a major histocompatibility complex-linked LMP gene. Nature 353: 664–667

McMichael AJ (1978) HLA restriction of human cytotoxic T lymphocytes specific for influenza virus: poor recognition of virus associated with HLA A2. J Exp Med 148: 1458
McMichael AJ, Askonas BA (1978) Influenza virus specific cytotoxic T cells in man; induction and properties of the cytotoxic T cell. Eur J Immunol 8: 705–710
McMichael AJ, Ting A, Zweerink HJ, Askonas BA (1977) HLA restriction of cell mediated lysis of influenza virus infected human cells. Nature 270: 524–526
McMichael AJ, Gotch FM, Cullen P, Askonas BA, Webster RG (1981). The human cytotoxic T cell response to influenza A vaccination. Clin Exp Immunol 43: 276–285
McMichael AJ, Gotch FM, Dongworth DW, Clark A, Potter CW (1938a) Declining T cell immunity to influenza 1977–1982. Lancet ii: 762–764
McMichael AJ, Gotch FM, Noble GR, Beare PAS (1983b) Cytotoxic T-cell immunity to influenza. N Engl J Med 309: 13–17
McMichael A, Gotch F, Rothbard J (1986) HLA B37 determines an influenza A virus nucleoprotein epitope recognized by cytotoxic T lymphocytes. J Exp Med 164(1397): 1397–1406
Milligan GN, Morrison LA, Gorka J, Braciale VL, Braciale TJ (1990) The recognition of a viral antigenic moiety by class I MHC-restricted cytolytic T lymphocytes is limited by the availability of the endogenously processed antigen. J Immunol 145(10): 3188–3193
Momburg F, Ortiz-Navarrete V, Neefjes J, Goulmy E, van de Wal Y, Spits H, Powis SJ, Butcher GW, Howard JC, Walden P, Hammerling GJ (1992) Proteasome subunits encoded by the major histocompatibility complex are not essential for antigen presentation. Nature 360: 174–177
Monaco J, Cho S, Attaya M (1990) Transport protein genes in the murine MHC: possible implications for antigen processing. Science 250: 1723–1726
Morris AG, Lin Y-L, Askonas BA (1982) Immune interferon release when a cloned cytotoxic T cell meets its correct influenza-infected target cell. Nature 295: 150–152
Morrison J, Elvin J, Latron F, Gotch F, Moots R, Strominger JL, McMichael AJ (1992) Identification of the nonamer peptide from influenza A matrix protein and the role of pockets of HLA A2 in its recognition by cytotoxic T lymphocytes. Eur J Immunol 22: 903–907
Morrison LA, Lukacher AE, Braciale VL, Fan D, Braciale TJ (1986) Differences in antigen presentation to MHC class I- and class II-restricted influenza virus-specific cytolytic T lymphocyte clones. J Exp Med 163: 903
Moss PAH, Moots RJ, Rosenberg WMC, Rowland-Jones SJ, McMichael AJ, and Bell JI (1991) Extensive conservation of alpha and beta chains of the human T cell antigen receptor recognizing HLA-A2 and influenza matrix peptide. Proc Natl Acad Sci USA 88: 8987–8991
Nathenson SG, Geliebter J, Pfaffenbach GM, Zeff RA (1986) Murine major histocompability complex class I mutants: molecular analysis and structure function implications. Annu Rev Immunol 4: 471–521
Nuchtern JG, Bonifacino JS, Biddison WE, Klausner RD (1989) Brefeldin A implicates egress from the endoplasmic reticulum in class I restricted antigen presentation. Nature 339: 223–226
Oehen S, Waldner H, Kundig TM, Hengartner H, Zinkernagel RM (1992) Antivirally protective cytotoxic T cell memory to lymphocytic choriomeningitis virus is governed by persisting antigen. J Exp Med 176(5): 1273–1281
Ortiz-Navarette V, Seelig A, Gernold M, Frentzel S, Kloetzel PM, Hammerling GJ (1991) Subunit of the 20S proteasome (multicatalytic proteinase) encoded by the major histocompatibility complex. Nature 353: 6662–6664
Palese P, Young JK (1983) Molecular epidemiology of influenza virus. In: Palese P, Kingsbury D (eds) Genetics of influenza virus. Springer, Vienna New York, pp 321–326
Parham P, Lomen CE, Lawlor DA, Ways JP, Holmes N, Coppin HL, Salter RD, Wan AM, Ennis PD (1988) Nature of polymorphism in HLA-A, -B and -C molecules. Proc Natl Acad Sci USA 85: 4005–4009
Parker KC, Bednarek MA, Hull LK, Utz U, Cunningham B, Zweerink HJ, Biddison WE, Coligan JE (1992) Sequence motifs important for peptide binding to the human MHC class I molecule, HLA-A2. J Immunol 149(11): 3580–3587
Plata F, Autran B, Martins LP, Wain-Hobson S, Raphael M, Mayaud C, Denis M, Guillon JM, Debre P (1987) AIDS virus specific cytotoxic T lymphocytes in lung disorders. Nature 328: 348–351
Robertson L, Caley JP, Moore J (1958) Importance of staphylococcus aureus in pneumonia in the 1957 epidemic of influenza A. Lancet ii: 233–236

Rotzschke O, Falk K (1991) Naturally-occurring peptide antigens derived from the MHC class-I-restricted processing pathway. Immunol Today 12(12): 447–455
Sethi KK, Naher H, & Stroehmann I (1988) Phenotypic heterogeneity of cerebrospinal fluid-derived cytotoxic T cell clones. Nature 335: 178–180
Silver ML, Guo H-C, Stominger JL, Wiley DC (1992) Atomic structure of a human MHC molecule presenting an influenza virus peptide. Nature 360: 367–369
Spies T, Bresnahan M, Bahram S, Arnold D, Blanck GEM, Pious D, DeMars R (1990) A gene in the major histocompatibility complex class II region controlling the class I antigen presentation pathway. Nature 348: 744–747
Spies T, Cerundolo V, Colonna M, Cresswell P, Townsend A, De Mars R (1992) Presentation of endogenous viral antigen dependent on putative peptide transporter heterodimer. Nature 355: 644–646
Sutton J, Rowland-Jones S, Rosenberg W, Nixon D, Gotch F, Gao M, Murray N, Spoonas A, Driscoll P, Smith M, Willis A, McMichael AJ (1993) A sequence pattern for peptides presented to cytotoxic T lymphocytes by HLA B8 revealed by analysis of epitopes and eluted peptides. Eur J Immunol 23: 447–453
Takahashi H, Takeshita T, Morein B, Putney S, Germain RN, Berzofsky JA (1990) Induction of $CD8^+$ cytotoxic T cells by immunization with purified HIV-1 envelope protein in ISCOMs [see comments]. Nature 344(6269): 873–875
Tite JP, Hughes JC, O'Callaghan D, Dougan G, Russell SM, Gao XM, Liew FY (1990) Anti-viral immunity induced by recombinant nucleoprotein of influenza A virus. II. Protection from influenza infection and mechanism of protection. Immunology 71 (2): 202–207
Townsend A, Bodmer H (1989) Antigen recognition by class I-restricted cytotoxic T lymphocytes. Annu Rev Immunol 7: 601–624
Townsend ARM, Skehel JJ (1982) Influenza A specific cytotoxic T cell clones that do not recognise viral glycoproteins. Nature 300: 655
Townsend ARM, McMichael AJ, Carter NP, Huddlestone JA, Brownlee GG (1984) Cytotoxic T cell recognition of the influenza nucleoprotein and haemagglutinin expressed in transfected mouse L cells. Cell 39: 13–25
Townsend ARM, Gotch FM, Davey J (1985) Cytotoxic T cells recognize fragments of the influenza nucleoprotein. Cell 42: 457–467
Townsend ARM, Bastin J, Gould K, Brownlee GG (1986a) Cytotoxic T lymphocytes recognise influenza haemagglutinin that lacks a signal sequence. Nature 234: 575
Townsend ARM, Rothbard J, Gotch FM, Bahadur G, Wraith D, McMichael AJ, (1986b) The epitopes of influenza nucleoprotein recognized by cytotoxic T lymphocytes can be defined with short synthetic peptides. Cell 44: 959–968
Townsend A, Ohlen C, Bastin J, Ljunggren HG, Foster L, Karre K (1989a) Association of class I major histocompatibility heavy and light chains induced by viral peptides. Nature 340(6233): 443–448
Townsend A, Ohlen C, Foster L, Bastin J, Ljunggren HG, Karre K (1989b) A mutant cell in which association of class I heavy and light chains is induced by viral peptides. Cold Spring Harb Symp Quant Biol 1: 299–308
Townsend A, Elliott T, Cerundolo V, Foster L, Barber B, Tse A (1990) Assembly of MHC class I molecules analyzed in vitro. Cell 62(2): 285–295
Trowsdale J, Hanson I, Mockridge I, Beck S, Townsend ARM, Kelly A (1990) Sequences encoded in the class II region of the MHC related to the ABC superfamily of transporters. Nature 348: 741–744
Ulmer J, Donnelly JJ, Parker SE, Rhodes GH, Felgner PL, Dwarki VJ, Gromkowski SH, Deck RR, DeWitt CM, Friedman A, Hawe LA, Leander KR, Martinez D, Perry HC, Shiver JW, Montgomery DL, Liu MA (1993) Heterologous protection against influenza by injection of DNA encoding a viral protein. Science 259: 1749–1749
Varghese JN, Laver WG, Colman PM, (1983) Structure of the influenza glycoprotein antigen neuraminidase at 2.9Å resolution. Nature 305: 35–40
Webster RG, Askonas BA (1980) Cross-protection and cross-reactive cytotoxic T-cells induced by influenza virus vaccines in mice. Eur J Immunol 10: 396–402
Webster RG, Laver WG, Air GM, Schild GC (1982) Molecular mechanisms of variation in influenza viruses. Nature 296: 115–121
Wells MA, Ennis FA, Albrecht P (1981a) Recovery from a viral repsiratory tract infection. 1. Influenza pneumonia in normal and T deficient mice. J Immunol 126: 1036–1041

Wells MA, Ennis FA, Albrecht P (1981b) Recovery from a viral respiratory infection. II. Passive transfer of immune spleen cells to mice with influenza pneumonia. J Immunol 126: 1042–1046
Wiley DC, Wilson IA, Skehel JJ (1981) Structural identification of the antibody-binding sites of Hong Kong influenza haemagglutinin and their involvement in antigenic variation. Nature 289: 373–378
Yap KL, Ada GL (1978a) Cytotoxic T cells in the lungs of mice infected with influenza A virus. Scand J Immunol 7: 73–80
Yap KL, Ada GL (1978b) Transfer of specific cytotoxic T lymphocytes protects mice inoculated with influenza virus. Nature 273: 238–240
Yewdell JW, Bennink JR, Smith GL, Moss B (1985) Influenza A virus nucleoprotein is a major target for cross-reactive anti-influenza A virus specific cytotoxic T lymphocytes. Proc Natl Acad Sci USA 82: 1785–1789
Yewdell YW, Bennink JR (1989) Brefeldin A specifically inhibits presentation of protein antigens to cytotoxic T lymphocytes. Science 244: 1072
Zhang W, Young AC, Imarai M, Nathenson SG, Sacchettini JC (1992) Crystal structure of the major histocompatibility complex class I H-2Kb molecule containing a single viral peptide: implications for peptide binding and T-cell receptor recognition. Proc Natl Acad Sci USA 89(17): 8403–8407
Zinkernagel RM, Doherty PC (1979). MHC-restricted cytotoxic T cells: studies on the biological role of polymorphic major transplantation antigens determining T cell restriction-specificity, function and responsiveness. Adv Immunol 27: 51
Zweerink HJ, Gammon MC, Utz U, Sauma SY, Harrer T, Hawkins JC, Johnson RP, Sirotina A, Hermes JD, Walker BD, et al. (1993) Presentation of endogenous peptides to MHC class I-restricted cytotoxic T lymphocytes in transport deletion mutant T2 cells. J Immunol 150(5): 1763–1771
Zweerink HT, Askonas BA, MIlican D, Courtneidge SA, Skehel JJ (1977) Cytotoxic T cells to type A influenza virus; viral haemagglutinin induces A-strain specificity while infected cells confer cross-reactive cytotoxicity. Eur J Immunol 7: 630–635

Cytotoxic T Lymphocytes in Dengue Virus Infection

I. KURANE and F. A. ENNIS

1 Introduction

1.1 Dengue Viruses

Dengue viruses are members of the family Flaviviridae and there are four serotypes, dengue virus types 1, 2, 3, and 4. The genome of dengue viruses consists of a single-stranded RNA nearly 11 kb in length which is plus-

Division of Infectious Diseases and Immunology, Department of Medicine, University of Massachusetts Medical Center, Worcester, MA 01655, USA

Current Topics in Microbiology and Immunology, Vol. 189

stranded and infectious (RICE et al. 1986). Dengue virus genome codes for three structural proteins, capsid (C), PreM, which is a precursor to membrane (M), and envelope (E), and seven nonstructural proteins, NS1, NS2a, NS2b, NS3, NS4a, NS4b, and NS5 (ZHAO et al. 1986; MACKOW et al. 1987; HENCHAL and PUTNAK 1990). The functions of these proteins are not clearly understood. NS3 protein has been reported to have protease activity (PREUGSHAT et al. 1990) and also contains conserved epitopes characteristic of helicases (GOLBALENYA et al. 1989). The virion consists of RNA genome surrounded by a lipid bilayer containing E and M proteins. Dengue viruses infect monocytes and other macrophages (HALSTEAD et al. 1977), B lymphoblastoid cells (SUNG et al. 1975; THEOFILOPOULOS et al. 1976), T cell tumor lines (KURANE et al. 1990), epithelial cells (SMITH and WRIGHT 1985), fibroblasts (KURANE et al. 1992), and endothelial cells (ANDREWS et al. 1978) in vitro. However, it is believed that in vivo monocytes and other macrophages are the important source of dengue virus replication (HALSTEAD et al. 1977; BOONPUCKNAVIG et al. 1976, 1979).

1.2 Illness Caused by Dengue Virus Infection

Dengue virus infection can be asymptomatic or cause two forms of illness, dengue fever (DF) and dengue hemorrhagic fever (DHF; HALSTEAD 1980, 1992; HAYES and GUBLER 1992). DF is a self-limited febrile disease. After an incubation period of 2–7 days, a sudden onset of fever occurs and body temperature rapidly rises to 39.4°C–41.1°C. The fever is usually accompanied by frontal or retroorbital headache. Back pain occasionally precedes fever. Myalgia or bone pain occurs soon after onset. A transient, macular, generalized rash which blanches under pressure as well as nausea, vomiting, lymphadenopathy and taste aberrations may develop. One to two days after defervescence a generalized morbilliform maculopapular rash appears, which spares the palms and soles and disappears in 1–5 days. Patients usually recover from these symptoms within 10 days after onset without complications.

On the other hand, some patients infected with dengue virus leak plasma into interstitial spaces, resulting in hypovolemia and possibly circulatory collapse. This severe and life-threatening syndrome, which is always accompanied by thrombocytopenia and sometimes frank hemorrhage, is termed DHF. The incubation period of DHF is presumed to be similar to that of DF. In the first phase, illness starts with relatively mild symptoms: a fever, malaise, vomiting, headache, anorexia, and cough. Rapid clinical deterioration and collapse follows after 2–5 days. In the second phase, the patients have cold clammy extremities, warm trunk, flushed face, restlessness, irritability, and midepigastric pain. This crisis lasts for 24–36 hours, and children recover rapidly once convalescene starts. The common hematologic abnormalities are an increase in hematocrit, thrombocytopenia, mild leukocytosis, prolonged bleeding time, and decreased prothrombin time. The World Health Organization categorizes DHF cases into four grades, from less severe (grade 1) to severe (grade 4; TECHNICAL ADVISORY COMMITTEE ON DENGUE HEMORRHAGIC FEVER FOR SOUTHEAST ASIAN AND WESTERN PACIFIC REGIONS 1980). Grades 3 and 4, in which plasma leakage is so profound that shock occurs, are also referred to as dengue shock syndrome (DSS).

1.3 Epidemiology of Dengue Virus Infection

Dengue viruses are transmitted to humans by mosquitoes, principally *Aedes aegypti.* Forest cycles of dengue virus transmission between monkeys and mosquitoes have also been reported (GUBLER 1988). Dengue virus infections are a serious cause of morbidity and mortality in many areas of the world: Southeast and South Asia, Australia, Central and South America, and the Caribbean (HALSTEAD 1981, 1988, 1992). Dengue virus infections can be estimated to occur in up to 100 million individuals yearly (HALSTEAD 1988). A total of 250 000–500 000 cases of DHF/DSS occur throughout the world annually, and case fatality rate is 1%–5% (HALSTEAD 1992). Since DHF/DSS was first recognized in the 1950s, 1.5 million cases have been hospitalized and 33 000 infected individuals have died as a result of this syndrome (HALSTEAD 1988). These facts indicate that dengue virus infection is one of the most important human infectious diseases.

It has been reported that in Thailand as many as 99% of cases of DHF/DSS occur in children who have antibody to dengue virus before dengue virus infection that causes DHF/DSS (SANGKAWIBHA et al. 1984; HALSTEAD 1980, 1988). These patients can be divided into two groups: (1) children who are over 1 year of age and show secondary antibody responses and (2) children who are less than 1 year of age and show primary immune responses. Children who are over 1 year of age comprise about 90% of all DHF/DSS cases. These cases are observed in secondary infections by a dengue virus of a different serotype from that which caused the primary infections. On the other hand, children who are less than 1 year of age and who develop DHF/DSS during primary infections are those born to dengue virus antibody-positive mothers. These observations have been supported by other epidemiological studies in Thailand and in Cuba (BURKE et al. 1988; GUZMAN et al. 1984, 1990). It has been reported that antibodies to dengue viruses at subneutralizing concentrations augment dengue virus infection of Fcγ receptor-positive cells such as monocytes (HALSTEAD and O'ROURKE 1977). Dengue virions and immunoglobulin G (IgG) antibody to dengue virus form virus-antibody complexes, and binding of these virus–antibody complexes to FcγR via the Fc portion of IgG results in augmentation of dengue virus infections (KURANE et al. 1991a). Based on the epidemiological and laboratory studies it has been hypothesized that antibodies to dengue viruses and other serotype cross-reactive-immune responses contribute to the pathogenesis of DHF/DSS. However, the pathogensis of DHF/DSS is not clearly understood.

2 Dengue Virus-Specific Human Cytotoxic T Lymphocytes

It is generally accepted that cytotoxic T lymphocytes (CTL) play a critical role during viral infections. Adoptive transfer of virus-specific (CTL) can protect mice from disease caused by lymphocytic choriomeningitis virus (LCMV) or influenza virus (BYRNE and OLDSTONE 1984; MOSKOPHIDIS et al. 1987; KUWANO et al. 1988). On the other hand, depending on timing and number of cells, the adoptive transfer of virus-specific CTL has resulted in development of significantly severe symptoms during infection with LCMV or respiratory syncytial virus (BAENZIGER et al. 1986; CANNON et al. 1988). These observations suggest that virus-specific CTL contribute to recovery from virus infections, but can also induce immunopathology in certain situations.

Analysis of CTL responses in dengue virus infections is important for understanding the mechanisms of recovery from dengue virus infections and the pathogenesis of DHF/DSS. This knowledge would be useful for the development of dengue vaccines which induce protective immune responses but do not induce immune responses that may lead to immunopathology. There is no good animal model to analyze DHF/DSS; therefore, analysis of dengue virus-specific human T lymphocytes is necessary to understand the role of T lymphocytes in dengue virus infections. In this review, we will focus on human cytotoxic T cell responses to dengue viruses and discuss possible roles of these CTL in the pathogenesis of DHF/DSS.

2.1 Dengue Virus-Specific Human $CD4^+$ $CD8^-$ CTL

2.1.1 Detection of $CD4^+$ CTL in Bulk Cultures

Dengue virus-specific human memory $CD4^+$ $CD8^-$ T lymphocytes were first demonstrated by KURANE et al. (1989 a) using peripheral blood mononuclear cells (PMBC) from healthy individuals who had been naturally infected with dengue viruses or immunized with a live, attenuated experimental vaccine. Dengue antigen-specific proliferation of these $CD4^+$ $CD8^-$ T memory cells was induced by noninfectious dengue antigens (Ag) prepared from infected green monkey kidney cells. $CD4^+$ memory T cells induced by primary dengue virus infection were mainly serotype-specific, but serotype-cross-reactive responses were noted at lower, but significant, levels (KURANE et al. 1989b). In some individuals, however, $CD4^+$ memory T cells after primary infections were solely serotype-specific (unpublished observation). Dengue virus-specific $CD4^+$ $CD8^-$ memory T lymphocytes stimulated by noninfectious dengue virus Ag for 7 days in bulk cultures lysed dengue Ag-pulsed autologous target cells in HLA class II-restricted fashion (unpublished observation). These cells did not lyse dengue Ag-pulsed autologous target cells without in vitro incubation. These observations suggest that dengue virus infections induce dengue virus-specific $CD4^+$ CTL.

2.1.2 Dengue Virus-Specific $CD4^+$ CTL Clones

Dengue virus-specific $CD4^+$ CTL clones were established and analyzed for virus and dengue serotype specificities (KURANE et al. 1991b). These CTL clones lysed dengue virus-infected or dengue Ag-pulsed autologous target cells. Thirteen dengue virus-specific $CD4^+$ CTL clones were established from an individual who had been immunized with yellow fever vaccine 2 years earlier and had been infected with dengue 3 virus 1 year before. In bulk culture proliferation assays, the PBMC from this donor responded to dengue 3 antigen, and they also responded to dengue 1, 2, and 4 antigens to lower, but significant, levels. Six patterns of virus and dengue serotype specificities were observed (Table 1). All of these $CD4^+$ CTL clones were HLA class II restricted, and HLA DP, HLA DQ, and HLA DR were all used as restriction elements by various clones (KURANE et al. 1991b). Dengue virus-specific $CD4^+$ CTL clones were also established from the PBMC of four other individuals (Table 1). $CD4^+$ CTL clones from each individual have at least

Table 1 Virus and dengue serotype specificity of dengue virus-specific cytotoxic T lymphocyte (CTL) clones

Donor (Type of infection)	Type of clone	Virus and dengue serotype specificity of clones
$CD4^+$ $CD8^-$ CTL		
# 1 (D3)	Serotype-specific	D3
	Subcomplex-specific	D1, D2, D3; D2, D3, D4
	Serotype-cross-reactive	D1, D2, D3, D4
	Flavi-cross-reactive	D1, D2, D3, D4, WNV
		D1, D2, D3, D4, WNV, YFV
# 2 (D4)	Serotype-specific	D4
	Subcomplex-specific	D2, D4
# 3 (D4)	Serotype-specific	D4
	Subcomplex-specific	D1, D3, D4
# 4 (D4)	Serotype-specific	D4
	Subcomplex-specific	D2, D4
# 5 (D1)	Serotype-specific	D1
	Subcomplex-specific	D1, D3
$CD8^+$ $CD4^-$ CTL		
# 2 (D4)	Serotype-specific	D4
	Subcomplex-specific	D2, D4
	Serotype-cross-reactive	D1, D2, D3, D4

D1 –4, dengue virus types 1–4; WNV, West Nile virus; YFV, yellow fever virus.

two patterns of virus and dengue serotype specificities. These results indicate that serotype specificities of dengue virus-specific $CD4^+$ CTL are heterogeneous in each individual and that infection with one serotype of dengue virus induces both serotype-specific and serotype-cross-reactive $CD4^+$ memory CTL in most individuals.

2.2 Dengue Virus-Specific $CD8^+$ $CD4^-$ CTL

Dengue virus-specific $CD8^+$ $CD4^-$ CTL were reported by BUKOWSKI et al. (1989), using PBMC from an individual who had been infected with dengue 4 virus. They incubated this donor's PBMC with infectious dengue viruses for 7–9 days and demonstrated that these PBMC lysed dengue virus-infected autologous fibroblasts. The effector cells were determined to be $CD8^+$ $CD4^-$ T cells and the lysis was HLA class I restricted. The $CD8^+$ CTL generated in bulk cultures were serotype cross-reactive. Dengue virus-specific $CD8^+$ CTL clones were established from the PBMC of the same individual. These clones had three patterns of serotype specificities: (1) specific for dengue virus type 4, (2) cross-reactive for dengue virus types 2 and 4, and (3) cross-reactive for dengue virus types 1, 2, 3, and 4 (Table 1). Although dengue virus-specific $CD8^+$ CTL have been analyzed using the PBMC from one individual so far, the results suggest that the dengue virus-specific $CD8^+$ memory CTL of an individual who was infected with one serotype of dengue virus are also heterogeneous, and both serotype-specific and serotype-cross-reactive $CD8^+$ memory CTL are induced after primary infection.

2.3 Proteins Recognized by Dengue Virus-Specific CTL

2.3.1 Proteins Recognized by $CD4^+$ CTL

As described above, dengue virus genes code for three structural proteins, C, PrM, E, and seven nonstructural proteins, NS1, NS2a, NS2b, NS3, NS4a, NS4b, and NS5. Proteins recognized by dengue virus-specific $CD4^+$ CTL have been determined mainly using dengue virus-specific clones. KURANE et al. (1991b) reported that seven of 12 dengue virus-specific $CD4^+$ clones recognized NS3 protein (Table 2). These included a dengue 3-specific clone, two dengue subcomplex-specific clones, three dengue serotype-cross-reactive clones, and a flavivirus cross-reactive clone. These results suggest that the NS3 protein has at least four epitopes recognized by the dengue virus-specific $CD4^+$ CTL of this donor. NS3 was also recognized by a $CD4^+$ T cell clone specific for dengue virus types 1, 3, and 4, established from a dengue 4-immune donor, and by a dengue 4-specific clone and a clone cross-reactive for dengue virus types 2 and 4 established from another dengue 4-immune donor (Table 2).

Table 2. Dengue virus proteins recognized by dengue virus-specific cytotoxic T lymphocytes (CTL)

Donor (Type of infection)	Protein recognized	Types of clones
$CD4^+$ $CD8^-$ CTL		
# 1 (D3)	NS3	Serotype-specific Subcomplex-specific Serotype-cross-reactive Flavi-cross-reactive
# 2 (D4)	NS3	Serotype-specific Subcomplex-specific
	E	Serotype-specific
# 3 (D4)	NS3	Subcomplex-specific
	E	Serotype-specific
	C or PrM	(Undetermined)
$CD8^+$ $CD4^-$ CTL		
# 2 (D4)	NS3	Serotype-specific Subcomplex-specific Serotype-cross-reactive
	E	(Bulk culture)

E protein was recognized by a dengue 4-specific clone from a dengue 4-immune donor, and by four dengue 4-specific clones from another dengue 4-immune donor (Table 2). Recently, we have established dengue virus-specific $CD4^+$ CTL clones from two individuals which recognize either C or PrM protein. These results indicate that NS3 and E proteins are the two major proteins recognized by dengue virus-specific $CD4^+$ CTL, and C or preM protein is also recognized by some $CD4^+$ CTL. These results suggest that multiple epitopes on NS3 protein are recognized by distinct $CD4^+$ CTL clones with a variety of virus and dengue serotype specificities and that epitope (s) on E protein are recognized by serotype-specific clones. However, more studies are needed to confirm these observations.

2.3.2 Proteins Recognized by $CD8^+$ CTL

Bukowski et al. (1989) have reported that $CD8^+$ CTL generated in bulk cultures from PBMC of a dengue 4 virus-immune donor recognized nonstructural proteins and E proteins, but did not identify which nonstructural proteins were recognized (Table 2). Recently, we established dengue virus-specific $CD8^+$ CTL clones from this individual and found that all of the $CD8^+$ CTL clones examined to date recognize epitopes on NS3 protein (Table 2). These clones include a dengue 4-specific clone, three clones cross-reactive for dengue virus types 2 and 4, and one clone cross-reactive for dengue virus types 1, 2, 3, and 4. These results indicate that NS3 and E proteins are recognized by the $CD8^+$ CTL of this donor and suggest that NS3 contains major $CD8^+$ CTL epitopes.

2.4 Epitopes Recognized by Dengue Virus-Specific CTL

The epitopes recognized by dengue virus-specific CTL have not been defined in detail; however, the epitopes have been localized using recombinant vaccinia viruses to some degree. The epitopes recognized by $CD4^+$ CTL clones established from a dengue 3 virus-immune donor were localized within amino acids (aa) 1–452 on the NS3 protein. There seems to be at least four $CD4^+$ CTL epitopes in this region, because $CD4^+$ CTL clones with four different serotype cross-reactivities recognize epitopes on aa 1–452.

Epitopes recognized by $CD8^+$ CTL clones have been localized within aa 453–618 on NS3 protein. These $CD8^+$ CTL clones have three different serotype cross-reactivities; therefore, there seems to be at least three $CD8^+$ CTL epitopes within aa 453–618 of NS3.

2.5 Activation of Dengue Virus-Specific CTL In Vivo

Dengue virus-specific $CD4^+$ $CD8^-$ and $CD4^-$ $CD8^+$ memory T cells are present in individuals who have been infected and recovered from dengue virus infections (KURANE et al. 1989a; BUKOWSKI et al. 1989). These T lymphocytes acquire CTL activity upon stimulation with dengue virus antigens in vitro, suggesting that these memory CTL are also activated and become cytotoxic in vivo. It is important to determine whether these dengue virus-specific CTL are activated and mediate cytotoxic T cell functions in vivo during dengue virus infections. It is known that activated T lymphocytes produce interleukin 2 (IL-2; SMITH 1980; BALKWILL and BURKE 1989) and interferon gamma (IFN-γ; GRANELLI-PIPERNO et al. 1986; BALKWILL and BURKE 1989) and release soluble IL-2 receptor (sIL-2R; RUBIN et al. 1985). $CD4^+$ T cells and $CD8^+$ T cells release soluble CD4 molecule (sCD4) and soluble CD8 molecule (sCD8) upon activation, respectively (FUJIMOTO et al. 1983; TOMKINSON et al. 1989). Therefore, levels of these markers in the sera should reflect the activation of T cells in vivo, although these markers do not reflect the cytotoxic functions of these T cells.

We recently reported that levels of IL-2, IFN-γ, sIL-2R, and sCD4 are elevated in the sera of patients with DHF/DSS or DF compared to the levels of these markers in the sera of healthy Thai children (KURANE et al. 1991c). The level of sCD8 was highly elevated in patients with DHF/DSS, but not in patients with DF. The levels of sIL-2R, sCD4, and sCD8 in the patients with DHF/DSS were significantly higher than the levels in patients with DF. These results strongly suggest that T lymphocytes are activated in vivo during dengue virus infection, and activation of $CD4^+$ and $CD8^+$ T cells is much greater in patients with DHF/DSS than in patients with DF. These results suggest that high levels of T cell activation may be associated with the pathogenesis of DHF/DSS. However, the possibility that the higher levels of $CD4^+$ and $CD8^+$ T cell activation in DHF/DSS may reflect a high level of inflammatory responses rather than actually causing the symptoms of DHF/DSS cannot be ruled out yet.

3 Analyses of Dengue Virus-Specific CTL Using Animal Models

Animal models are not available to analyze the complications of severe dengue virus infections, DHF/DSS, although development of animal models for dengue virus research has been attempted. However, mice have been successfully used to analyze dengue virus-specific CTL. Dengue virus-specific $CD8^+$ CTL were demonstrated by PANG et al. (1988) using dengue virus-infected mice. ROTHMAN et al. demonstrated dengue virus-specific $CD8^+$ CTL using mice with H-2^b, H-2^d, or H-2^k haplotypes, and dengue virus proteins recognized by CTL generated in bulk cultures were determined. Interestingly, CTL from mice with these three haplotypes show different patterns of protein recognition. $CD8^+$ CTL from H-2^k mice predominantly recognize NS3 protein, and those from H-2^b mice recognize NS2b, NS4a, or NS4b protein. $CD8^+$ CTL from H-2^d mice recognize NS3, structural proteins, and NS1 or NS2a protein (A. L. Rothman, unpublished observations). This heterogeneity of protein recognition depending on major histocompatibility complex (MHC) is consistent with observations of dengue virus-specific human CTL. However, it is of interest that NS3 protein is recognized by murine CTL from two of the three H-2 haplotypes. Analyses of dengue virus-specific $CD4^+$ CTL using mouse models have not been reported yet.

4 Possible Role of Dengue Virus-Specific CTL in Dengue Virus Infections

The role of T lymphocytes in dengue virus infections is not well understood. Based on the analysis of CTL in other virus systems, such as influenza virus and LCMV, and on the observation that dengue virus-specific $CD4^+$ and $CD8^+$ CTL lyse dengue virus-infected autologous cells, it is reasonable to consider that dengue virus-specific CTL eliminate dengue virus-infected monocytes, thereby controlling dengue virus infections.

Possible immunopathological roles for serotype-cross-reactive immune responses in dengue virus infections have been suggested based on epidemiological observations. Most patients with DHF/DSS recover quickly after receiving intravenous fluids, and little pathologic changes have been observed compared to the severity of disease in patients with DHF/DSS. Perivascular edema was observed, but destruction of vascular endothelial cells was not apparent (BAHMARAPRAVATI et al. 1968). These clinical and pathological studies suggest that plasma leakage observed in DHF/DSS is due to malfunction of vascular endothelial cells rather than structural destruction. It is not known how dengue virus-specific $CD4^+$ and $CD8^+$ CTL contribute to the pathogenesis of DHF/DSS. However, the high levels of activation of $CD4^+$ and $CD8^+$ T cells in DHF/DSS, as demonstrated by the high sCD4 and sCD8 levels in the sera of patients with DHF/DSS, suggest possible roles for CTL in this immunopathology (KURANE and ENNIS 1992).

Antibodies to dengue virus enhance dengue virus infection of monocytes and other macrophages by forming dengue virus-antibody complexes during secondary infections with a serotype different from that which caused primary infection. Dengue virus-specific $CD4^+$ CTL are stimulated and produce IFN-γ, IL-2, and other lymphokines. IFN-γ upregulates the expression of FcγR on monocytes (GUYRE et al. 1983; PERUSSIA et al. 1983) and further augments antibody-dependent enhancement of infection (KONTNY et al. 1988). IFN-γ also upregulates the expression of HLA class I and class II molecules (Kelley et al. 1983). An increased number of dengue virus-infected cells and upregulation of HLA molecules facilitates recognition of dengue virus antigens by $CD4^+$ and $CD8^+$ T lymphocytes and results in much higher levels of T cell activation. Higher levels of T cell activation results in higher levels of lymphokines. IFN-γ-activated monocytes secrete higher levels of monokines and chemical mediators. Lysis of activated, dengue virus-infected monocytes may result in release of intracellular contents. These monocytes may release monokines and chemical mediators as the result of interaction with CTL. These interactions between CTL and infected monocytes could result in high levels of lymphokines, monokines, chemical mediators, and anaphylatoxins during a short period of time. A rapid increase in the levels of these mediators generated by interactions of dengue virus-infected monocytes and dengue virus-specific $CD4^+$ and $CD8^+$ CTL would lead to plasma leakage, shock, derangement of coagulation, and hemorrhagic manifestations of DHF/DSS. However, these possible mechanisms are not proven and future studies are necessary.

5 Future Research

As reviewed above, dengue virus-specific human CTL have been detected and analyzed to some degree. There are, however, many questions that remain to be answered. In this section we will discuss some of these important questions.

5.1 Determination of the Epitopes Recognized by Dengue Virus-Specific CTL

Recognition of proteins by dengue virus-specific CTL has been analyzed, and NS3 and E proteins seem to be two major proteins which contain multiple epitopes for $CD4^+$and $CD8^+$ CTL. However, the epitopes have not been defined at amino acid levels yet. Epitopes will be mapped by using overlapping synthetic peptides or by isolation and analysis of natural peptides from antigen-presenting cells (FALK et al. 1991; RUDENSKY et al. 1991). Determination of predominant CTL epitopes in conjunction with the definition of HLA alleles which present these epitopes will provide useful information for the development of effective subunit vaccines.

5.2 T Cell Receptor Usages of Dengue Virus-Specific CTL

It is understood that T cells recognize a peptide which bound to an HLA molecules via T cell antigen receptors (TcR). Past studies have suggested that T cells which express restricted sets of TcR-variable region genes are involved in the pathogenesis of certain inflammatory diseases (ACHA-ORBEA et al. 1988; WACHERPFENNIG et al. 1990). Selective clonal expansion of T cells with certain TcR may lead to pathologic conditions in DHF/DSS. It has been reported that suppression of T cells expressing certain TcR-variable region genes prevents T cell-mediated diseases (ACHA-ORBEA et al. 1988; URBAN et al. 1988). If dengue virus-specific CTL contribute to the pathogenesis of DHF/DSS, and these CTL express restricted sets of TcR genes, prevention and control of DHF/DSS may be possible. If dengue virus-specific CTL mainly contribute to recovery from infection, stimulation of T cells expressing certain sets of TcR may enhance protective immunity against dengue virus infection. Thus, analysis of T cell receptor usage of dengue virus-specific CTL will expand our understanding of human T cell responses to dengue viruses.

5.3 Lymphokine Production by Dengue Virus-Specific CTL

Recently, it has been reported that human $CD4^+$ T cells can be divided into two functionally distinct subsets similar to murine $CD4^+$ T cells (PELTZ 1991; ROMAGNANI 1991). Th1 cells produce IL-2, IFN-γ, and lymphotoxin upon stimulation, whereas Th2 cells produce IL-4, IL-5, IL-6, and IL-10. In inflammatory and allergic diseases, one subset becomes dominant, and it is assumed that the dominant subset of $CD4^+$ T cells contribute to the pathogenesis by producing a unique set of lymphokines (DEL PRETE et al. 1991). It is also known that $CD8^+$ T cells have the ability to produce certain lymphokines as well as $CD4^+$ T cells (SALGAME et al. 1991). Although dengue virus-specific $CD4^+$ and $CD8^+$ CTL are mainly selected based on their cytotoxic activities, it is possible that these CTL contribute to the pathogenesis and recovery by their other functions such as lymphokine production. Analysis of lymphokine production by these dengue virus-specific CTL is important for our understanding of the roles of these CTL in vivo.

5.4 Animal Models to Analyze Dengue Virus Infection

The development of animal models to analyze the pathogenesis of systemic dengue virus infections has not been successful, although dengue virus-specific murine CTL were detected. Lethal intracerebral infection of mice with neuroadapted dengue virus has been used as a model to examine protective immunity against dengue virus infection (KAUFMAN et al. 1987; SCHLESINGER et al. 1987). One highly neuroadapted strain of dengue 1 virus induces clinical disease after intraperitoneal inoculation; however, disease manifests as encephalitis (HOTTA et al. 1981). These models are not sufficient, because encephalitis is not commonly seen in human dengue virus infections. The protective and immunopathological roles of dengue virus-specific CTL should be determined in vivo using reliable animal models. Successful development of animal models suitable for analyzing systemic dengue virus infections is needed.

5.5 Analyses of Dengue Virus-Specific CTL Activated In Vivo During Dengue Virus Infections

So far, dengue virus-specific human CTL have been analyzed using lymphocytes of individuals who had been infected with dengue virus and recovered from infection. Thus, analysis has been done by in vitro expansion and activation of memory CTL. It is important to understand the CTL responses in vivo during acute dengue virus infection. It has been reported that activated T lymphocytes are present in the PBMC of patients with DHF/DSS (KURANE et al. 1991c). Analysis of dengue virus-specific CTL using lymphocytes from patients with DF or DHF/DSS will be needed to further understand the role of these CTL in vivo.

6 Conclusions

Dengue virus infections are a major public health problem in tropical and subtropical areas of the world. The world Health Organization has stressed the importance of developing dengue virus vaccines (BRANDT 1988); however, safe and effective vaccines have not been developed. Dengue vaccines should develop protective immunity, but should not induce immune responses which may lead to DHF/DSS during future dengue virus infections. Although dengue virus-specific $CD4^+$ $CD8^-$ and $CD4^-$ $CD8^+$ CTL have been demonstrated and analyzed to some degree, there are many questions that should be addressed. The dengue virus proteins which contain CTL epitopes should be determined, and major CTL epitopes should be mapped. Furthermore, the role of these CTL in vivo needs to be clearly understood. Understanding human CTL response to dengue virus will provide useful information not only for the development of safe and effective vaccines, but also for more effective treatment of dengue virus infections.

Acknowledgements. This work was supported by grants from the U.S. Army Medical Research and Development Command (DAMA 17-86-C-6208) and from the National Institutes of Health (NIH-RO1-AI30624, NIH-T32-A107272). The opinions contained herein are those of the authors and should not be construed as representing the official policies of the Department of Army or the Department of Defense of the USA.

References

Acha-Orbea H, Mitchell DJ, Timmermann L, Wraith D, Tausch G, Waldor M, Zamvil S, McDevitt H, Steinman L (1988) Limited heterogeneity of T cell receptors from lymphocytes mediating autoimmune encephalomyelitis allows specific immune intervention. Cell 54: 263–273

Andrews BS, Theofilopoulos AN, Peters CJ, Loskutoff DJ, Brandt WE, Dixon FJ (1978) Replication of dengue and junin viruses in cultured rabbit and human endothelial cells. Infect Immun 20: 776–781

Baenziger J, Hengartner H, Zinkernagel RM, Cole GA (1986) Induction or prevention of immunopathological disease by cloned cytotoxic T cell lines specific for lymphocytic choriomeningitis virus. Eur J Immunol 16: 387–393

Balkwill FR, Burke F (1989) The cytokine network. Immunol Today 10: 299–304

Bhamarapravati N, Tuchinda P, Boonyapaknavik V (1968) Pathology of Thailand haemorrhagic fever: a study of 100 autopsy cases. Ann Trop Med Parasitol 61: 500–510

Boonpucknavig V, Bhamarapravati N, Boonpucknavig S, Futrakul P, Tanpaichitr P (1976) Glomerular changes in dengue hemorrhagic fever. Arch Pathol Lab Med 100: 206–212

Boonpucknavig S, Boonpucknavig V, Bhamarapravati S, Nimmanitya S (1979) Immunofluorescence study of skin rash in patients with dengue hemorrhagic fever. Arch Pathol Lab Med 103: 463–466

Brandt WE (1988) Current approaches to the development of dengue vaccines and related aspects of the molecular biology of flaviviruses. J Infect Dis 157: 1105–1111

Bukowski JF, Kurane I, Lai C-J, Bray M, Falgout B, Ennis FA (1989) Dengue virus-specific cross-reactive $CD8^+$ human cytotoxic T lymphocytes. J Virol 63: 5086–5091

Burke DS, Nisalak A, Johnson D, Scott RM (1988) A prospective study of dengue infections in Bangkok. Am J Trop Med Hyg 38: 172–180

Byrne JA, Oldstone MBA (1984) Biology of cloned cytotoxic T lymphocytes specific for lymphocytic choriomeningitis virus: clearance of virus in vivo. Virology 53: 682–686

Cannon MJ, Openshaw PJM, Askonas BA (1988) Cytotoxic T cells clear virus but augment lung pathology in mice infected with respiratory syncytial virus. J Exp Med 168: 1163–1168

Del Prete GF, De Carli M, Mastromauro C, Biagiotti R, Macchia D, Falagiani P, Ricci M, Romagnani S (1991) Purified protein derivative of mycoplasm tuberculosis and excretory-secretory antigen(s) of Toxocara canis expand in vitro human T cells with stable and opposite (type 1 T helper of type 2 T helper) profile of cytokine production. J Clin Invest 88: 346–350

Falk K, Rotzschke O, Stevanovic S, Jung G, Rammensee HG (1991) Allele-specific motifs revealed by sequencing of self-peptides eluted from MHC molecules. Nature 351: 290–296

Fujimoto J, Levy S, Levy R (1983) Spontaneous release of the Leu-2 (T8) molecule from human T cells. J Exp Med 159: 752–766

Golbalenya AE, Donchenko AP, Koonin EV, Blinov VM (1989) N-terminal domains of putative helicases of flavi- and pestiviruses may be serine proteases. Nucleic Acids Res 17: 3889–3897

Granelli-Piperno A, Andrus L, Steinman RM (1986) Lymphokine and nonlymphokine mRNA levels in stimulated human T cells. J Exp Med 163: 922–937

Gubler DJ (1988) Dengue PP223–260. In: Monath TP (ed) The arboviruses: epidemiology and ecology. CRC Press, Boca Raton

Guzman MG, Kouri GP, Bravo J, Soler M, Vazquez S, Santos M, Villaescusa R, Basanta P, Indan G, Ballester JM (1984) Dengue haemorrhagic fever in Cuba. II. Clinical investigations. Trans R Soc Trop Med Hyg 78: 239–241

Guzman MG, Kouri GP, Bravo J, Soler M, Vazquez S, Morier L (1990) Dengue hemorrhagic fever in Cuba, 1981: a prospective seroepidemiologic study. Am J Trop Med Hyg 42: 279–284

Guyre PM, Morganerli P, Miller R (1983) Recombinant immune interferon increases immunoglobulin G Fc receptors on cultured human mononuclear phagocytes. J Clin Invest 72: 393–397

Halstead SB (1980) Immunological parameters of togavirus diesease syndromes. Schlesinger RW (ed) The togaviruses: biology, structure, replication. Academic, New York, pp 107–173

Halstead SB (1981) Dengue hemorrhagic fever—a public health problem and a field for research. Bull WHO 58: 1–21

Halstead SB (1988) Pathogenesis of dengue. Challenges to molecular biology. Science 239: 476–481

Halstead SB (1992) Dengue viruses. In: Gorbach SL, Bartlett JG, Blacklow NR (eds) Infectious diseases. Saunders, Philadelphia, PP 1830–1835

Halstead SB, O'Rourke EJ (1977) Dengue viruses and mononuclear phagocytes. I. Infection enhancement by non-neutralizing antibody. J Exp Med 146: 210–217

Halstead SB, O' Rourke EJ, Allison AC (1977) Dengue virus and mononuclear phagocytes. II. Identity of blood and tissue leukocytes supporting in vitro infection. J Exp Med 146: 218–219

Hayes EB, Gubler DJ (1992) Dengue an dengue hemorrhagic fever. Pediatr Infect Dis J 11: 311–317

Henchal EA, Putnak R (1990) The dengue viruses. Clin Microbiol Rev. 3: 376–396

Hotta H, Murakami I, Miyasaki K, Takeda Y, Shirane H, Hotta S (1981) Inoculation of dengue virus into nude mice. J Gen Virol 52: 71–76

Kaufman BM, Summers PL, Dubois DR, Eckels KH (1987) Monoclonal antibodies against dengue 2 virus E-glycoprotein protect mice against lethal dengue infection. Am J Trop Med Hyg 36: 427–434

Kelley VE, Fiers W, Strom TB (1983) Cloned human interferon-γ, but not interferon-β or -α, induces expression of HLA-DR determinants by fetal monocytes and myeloid leukemic cell lines. J Immunol 132: 240–245

Kontny U, Kurane I, Ennis FA (1988) Gamma interferon augments Fc receptor-mediated dengue virus infection of human monocytic cells. J Virol 62: 3928–3933

Kurane I, Ennis FA (1992) Immunity and immunopathology in dengue virus infections. Semin Immunol 4: 121–127

Kurane I, Innis B, Nisalak A, Hoke C, Nimmanritya S, Meager A, Ennis FA (1989a) Human T cell responses to dengue virus antigens. Proliferative responses and interferon gamma production . J Clin Invest 83: 506–513

Kurane I, Meager A, Ennis FA (1989b) Dengue virus-specific human T cell clones: serotype cross-reactive proliferation, interferon gamma production, and cytotoxic activity. J Exp Med 170: 763–775

Kurane I, Kontny U, Janus J, Ennis FA (1990) Dengue-2 virus infection of human mononuclear cell lines and establishment of persistent infections. Arch Virol 110: 91–101

Kurane I, Mady BJ, Ennis FA (1991a) Antibody-dependent enhancement of dengue virus infection. Rev Med Virol 1: 211–221

Kurane I, Brinton MA, Samson AL, Ennis FA (1991b) Dengue virus-specific, human $CD4^+$

CD8⁻ cytotoxic T cell clones: multiple patterns of virus crossreactivity recognized by NS3-specific T cell clones. J Virol 65: 1823–1828
Kurane I, Innis BL, Nimmannitya S, Nisalak A, Meager A, Janus J, Ennis FA (1991c) Activation of T lymphocytes in dengue virus infections: high levels of soluble interleukin 2 receptor, soluble CD4, soluble CD8, interleukin 2 and interferon gamma in sera of children with dengue. J Clin Invest 88: 1473–1480
Kurane I, Janus J, Ennis FA (1992) Dengue virus infection of human spin fibroblasts in vitro: production of IFN-β, IL-6 and GM-CSF. Arch Virol 124: 21–30
Kuwano K, Scott M, Young JF, Ennis FA (1988) HA2 subunit of influenza A H1 and H2 subtype viruses induces a protective cross-reactive cytotoxic T lymphocyte response. J Immunol 140: 1264–1268
Mackow E, Makino Y, Zhao B, Zhang Y-M, Markoff L, Buckler-White A, Guiler M, Chanock R, Lai C-J (1987) The nucleotide sequence of dengue type 4 virus. Analysis of genes coding for nonstructural proteins. Virology 159: 217–228
Moskophidis D, Cobbold SP, Waldmann H, Lehmann-Grube F (1987) Mechanism of recovery from acute virus infection: treatment of lymphocytic choriomeningitis virus-infected mice with monoclonal antibodies reveals that Lyt-2+ T lymphocytes mediate clearance of virus and regulate the antiviral antibody response. J Virol 61: 1867–1874
Pang T, Devi S, Blanden RV, Lam SK (1988) T cell-mediated cytotoxicity against dengue-infected target cells. Microbiol Immunol 32: 511–518
Peltz G (1991) A role for CD4⁺ T-cell subsets producing a selective pattern of lymphokines in the pathogenesis of human chronic inflammatory and allergic diseases. Immunol Rev 123: 23–35
Perussia B, Dayton ET, Lazarus R, Fanning V, Trinchieri G (1983) Immune interferon induces the receptor for monomeric IgG1 on human monocytic and myeloid cells. J Exp med 158: 1092–1113
Preugschat F, Yao C-W, Strauss JH (1990) In vitro processing of dengue virus type 2 nonstructural proteins NS2A, NS2B and NS3. J Virol 64: 4364–4374
Rice CM, Strauss EG, Strauss JH (1986) Structure of the flavivirus genome. In: Schlesinger S, Schlesinger MJ (eds) The togaviridae and flaviviridae. Plenum, New York, pp 279–326
Romagnani S (1991) Human Th1 and Th2 subsets: doubt no more. Immunol Today 12: 256–257
Rubin LA, Kurman CC, Fritz ME, Biddison WE, Boutin B, Yarchoan B, Nelson DL (1985) Soluble interleukin 2 receptors are released from activated human lymphoid cells in vitro. J Immunol 135: 3172–3177
Rudensky AY, Preston-Hurlburt P, Hong S-C, Barlow A, Janeway CA (1991) Sequence analysis of peptides bound to MHC class II molecues. Nature 353: 622–627
Salgame P, Abrams JS, Clayberger C, Goldstein H, Convit J, Modlin RL, Bloom BR (1991) Differing lymphokine profiles of functional subsets of human CD4 and CD8 T cell clones. Science 254: 279–282
Sangkawibha N, Rojanasuphot S, Ahandrink S, Viriyapongse S, Jatanasen S, Salitul V, Phanthumachinda B, Halstead SB (1984) Risk factors in dengue shock syndrome: a prospective epidemiologic study in Rayong, Thailand. Am J Epidemiol 120: 653–669
Schlesinger JJ, Brandriss MW, Walsh EE (1987) Protection of mice against dengue 2 virus encephalitis by immunization with the dengue 2 virus non-structural glycoprotein NS1. J Gen Virol 68: 853–857
Smith KA (1980) T cell growth factor. Immunol Rev 31: 337–352
Smith GW, Wright PJ (1985) Synthesis of proteins and glycoproteins in dengue type 2 virus-infected Vero and Aedes albopictus cells. J Gen Virol 66: 559–571
Sung JS, Diwan AR, Falkler WA Jr, Yang H-Y, Halstead SB (1975) Dengue carrier culture and antigen production in human lymphoblastoid lines. Intervirology 5: 137–149
Technical Advisory Committee on dengue hemorrhagic fever for Southeast Asian and Western Pacific regions (1980) Guide for diagnosis, treatment and control of dengue hemorrhagic fever. World Health Organization, Geneva
Theofilopoulos AN, Brandt WE, Russell PK, Dixon FT (1976) Replication of dengue-2 virus in cultured human lymphoblastoid cells and subpopulations of human peripheral leukocytes. J Immunol 117: 953–961
Tomkinson BE, Brown MC, Ip SH, Carrabis S, Sullivan JL (1989) Soluble CD8 during T cell activation. J Immunol 142: 2230–2236

Urban JL, Kuman V, Kono DH, Gomez C, Horvath S, Clayton J, Ando D, Sercarz E, Hood L (1988) Restricted use of T cell receptor V genes in murine autoimmune encephalomyelitis raises possibilities for antibody therapy. Cell 54: 577–592

Wacherpfennig KW, Ota K, Endo N, Seidman J, Rosenzweig A, Weiner H, Hafler D (1990) Shared human T cell receptor Vβ usage to immunodominant regions of myelin basic protein. Science 248: 1016–1017

Zhao B, Mackow E, Buckler-White A, Markoff L, Chanock R, Lai C-J, Makino Y (1986) Cloning full-length dengue type 4 viral DNA sequences: analysis of genes coding for structural proteins. Virology 155: 77–88

Cytotoxic T Cells in Paramyxovirus Infection of Humans

S. DHIB-JALBUT[1] and S. JACOBSON[2]

1 Introduction

Paramyxoviruses include a number of human and animal pathogens that enter the host and spread through the respiratory tract. In humans, paramyxoviruses include measles virus (MV), mumps virus (MPS), parainfluenza virus (the causative agent of croup), and respiratory syncytial virus (RSV), which causes bronchitis and pneumonitis in infants. In animals, important pathogens include canine distemper, rinderpest, Newcastle disease, and parainfluenza viruses.

Paramyxoviruses are enveloped viruses with a diameter ranging from 150–250 nm. The envelope is a spherical lipoprotein that consists of the hemagglutinin (HA)–neuraminidase (HN) and the fusion (F) proteins. The core is a helical ribonucleoprotein which contains nonsegmented single-stranded RNA genome with a negative polarity. MV, which is the focus of this review, contains a 16-kb genome that codes for six structural proteins—nucleocapsid (NC), phosphoprotein (P), matrix (M), F, HA, large (L) protein—and at least two nonstructural viral proteins (C and V). The first five structural

[1] Neurology and Pathology Departments, University of Maryland at Baltimore, Baltimore, MD 21201, USA
[2] Neuroimmunology Branch, National Institutes of Neurological Disorders and Stroke, National Institutes of Health, Bethesda, MD, USA

Current Topics in Microbiology and Immunology, Vol. 189

proteins have been purified from an MV-infected Daudi cell line by affinity chromatography and from recombinant vaccinia viruses and have been used to study the cellular immune response to the individual viral proteins. The virus attaches to the host cell surface by binding of the HA protein to cell surface receptors followed by fusion of the virion envelope with the cell membrane. The role of neuraminidase and cleavage of the F protein in virus–cell interaction varies among the different members of the paramyxovirus family. Once in the cell cytoplasm, the viral transcriptase carried in the virion is used to synthesize viral mRNA complementary to the viral-negative stranded RNA. Viral proteins are then translated and assembled into virion that bud out from the host cell surface (KINGSBURY 1985).

This chapter reviews the cytotoxic T lymphocyte (CTL) response to three human paramyxoviruses, namely, MV, MPS, and RSV. The CTL response to measles virus has been more extensively studied than other paramyxoviruses and, therefore, it will be discussed in detail.

2 The CTL Response to Measles Virus

Measles virus is a contagious human pathogen that causes an acute exanthematous febrile illness which is self-limiting. In rare cases, serious complications involving the central nervous system (CNS) can take place. These include post-measles encephalomyelitis (PME), subacute sclerosing panencephalitis (SSPE), and subacute measles encephalitis (SME) in the immunocompromised host. Vaccination with the live-attenuated virus has resulted in near eradication of measles in the developed world and has reduced drastically the incidence of complications (NORRBY 1985).

A cellular immune response to MV develops during acute infection which is believed to be responsible for the exanthematous rash (SURINGA et al. 1970). Because of its lymphotropism, transient immunosuppression can occur during acute measles (NORRBY 1985). However, during convalescence a cytotoxic T cell response develops which is usually maintained over the life of the individual (KRETH et al. 1979; JACOBSON et al. 1985). Abnormalities in the CTL to MV have been found in multiple sclerosis (MS; JACOBSON et al. 1985), SSPE (DHIB-JALBUT et al. 1989a), and healthy individuals who had acquired immunity to MV through vaccination rather than natural infection (WU et al. 1993). Thus, we will discuss the MV CTL in relation to acute infection, immunity following natural infection, vaccination, and in disease states such as PME, SSPE, and MS.

2.1 Measles Virus CTL in Healthy Individuals

Measles infection produces a lifelong immunity, and neutralizing antibodies remain detectable throughout the life of the individual (NORRBY 1985). MV

CTL is not detectable in freshly isolated peripheral blood lymphocytes (PBL) from healthy individuals, but can be generated following in vitro stimulation with the virus (LUCAS et al. 1983; JACOBSON et al. 1985). PBL stimulated with the Edmonston strain of MV and cultured for 1 week generate CTL that lyse MV-infected Epstein-Barr virus (EBV)-transformed B cell targets. This response is measles specific, primarily CD4$^+$, and restricted by MHC class II molecules (JACOBSON et al. 1984, 1985, 1987). The magnitude of the MV CTL response in healthy individuals ranges between 18% and 50% with a mean of 26.0 ± 10.4% (*n*, 22) specific lysis (JACOBSON et al. 1985). MV-reactive CTL clones of either the CD4$^+$ or CD8$^+$ phenotype have been generated from healthy individuals with a remote history of measles infection (JACOBSON et al. 1984; VAN BINNENDIJK et al. 1989). The phenotype of the CTL clones generated appears to have a relationship to the antigen-presenting cell. Measles antigen presented by macrophages predominantly produces CD4$^+$ class II-restricted CTL clones, whereas measles antigen presented by EBV-transformed B cell predominantly stimulates the production of CD8$^+$ class I-restricted CTL clones (VAN BINNENDIJK et al. 1989).

In vitro stimulation of PBL with the MV polypeptides HA, F, NC, or M results in the generation of MV CTL responses (Fig. 1; JACOBSON et al. 1987; DHIB-JALBUT et al. 1989b), indicating that an exogenous endocytic pathway, in addition to an endogenous intracytoplasmic pathway, are used in processing MV antigen. In addition, CTL clones generated from cultures stimulated with the whole virus show diverse reactivity to different MV polypeptides including HA, F, and NC (VAN BINNENDIJK et al. 1989). The frequency of CD4$^+$ clones specific for F and NC seems to be higher than that for the HA protein. Vaccinia virus recombinants expressing HA, F, or NC have been shown to generate cellular immune responses and to be protective against

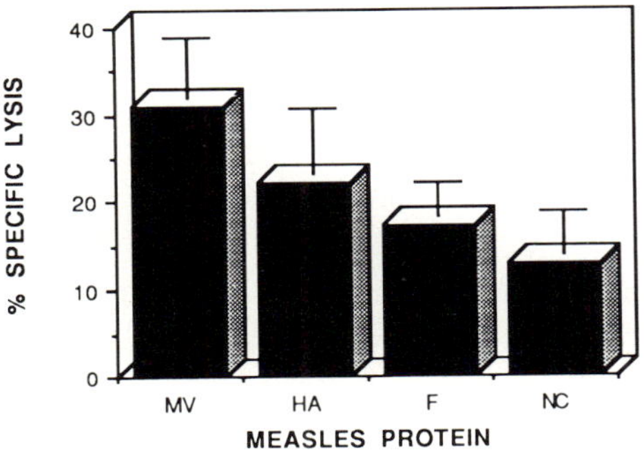

Fig. 1. Cytotoxic T lymphocyte response to measles virus (*MV*) and the MV proteins hemagglutinin (*HA*), fusion protein (*F*), and nucleocapsid (*NC*). Peripheral blood lymphocytes (PBL) were stimulated with MV or purified MV polypeptide and tested for lysis of MV-infected autologous Epstein-Barr virus-transformed B cells at an effector-to-target ratio of 20:1. The graph shows the mean (± standard deviation) percentage of specific lysis from three healthy individuals (DHIB-JALBUT et al. 1989a)

measles infection in mice and rats, which suggests a potential for MV subunit vaccine in humans (BRINKMANN et al. 1991).

The significance of MV CTL precursors in the production of long-term immunity to measles is not understood. However, recent studies in our laboratory of in vitro MV CTL response (which is predominantly $CD4^+$) in patients with vaccine failure and healthy vaccinated peer controls suggest a role for MV CTL in protection (WU et al. 1993). While the majority of patients with measles vaccine failure showed a high MV CTL response during convalesccence, vaccinated healthy peer controls had very low MV CTL compared to healthy individuals with a remote history of measles infection, but their influenza CTL was not different from the other groups (Fig. 2). This would seem to indicate that natural infection results in a higher and sustained MV CTL response compared to a low and perhaps unsustained MV CTL response in vaccinees, which puts the latter group at a risk for infection. This would suggest that $CD4^+$ MV CTL may be involved in the maintenance of long-term immunity. Support for a $CD4^+$ T cell role in protection against MV can be derived indirectly from knowledge of measles infection in children with human immunodeficiency virus (HIV) and experimental measles infection in Lewis Rats: children with HIV and low $CD4^+$ T cell counts develop fulminating measles pneumonitis, but survive acute measles if they have been vaccinated (KAPLAN et al. 1992). This would suggest that while $CD4^+$ T cells may not be required for disease production, they assist in reducing complications and enhancing recovery. Since measles antibody levels do not correlate with $CD4^+$ cell counts in asymptomatic HIV patients (SHA et al. 1991), it is possible that $CD4^+$ T-helper cells are sufficient to maintain measles antibodies, but reduction in $CD4^+$ MV CTL may increase the risk of complications

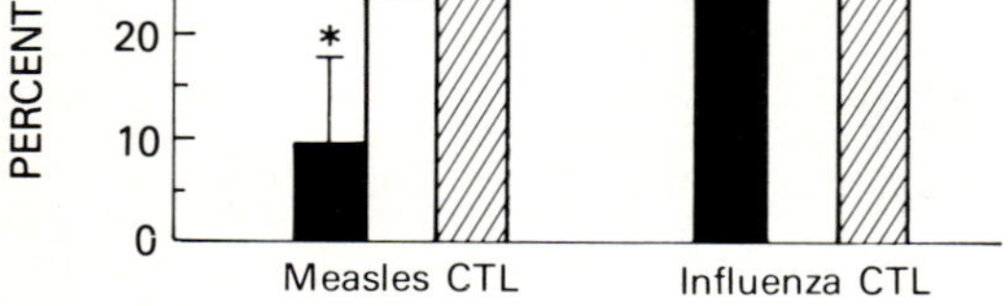

Fig. 2. Measles virus- and influenza virus-specific cytotoxic T lymphocyte responses in healthy, previously vaccinated individuals (vaccinated peers; $n = 7$), healthy adults with a history of natural infection ($n = 5$), and in patients 4–8 weeks after acute measles ($n = 18$). The graph shows the mean percentage of specific lysis ($\pm$ SD) of autologous virus-infected Epstein-Barr virus-transformed B cell targets at an effector-to-target ratio of 40:1. The *asterisk* indicates a statistically significant difference compared to healthy adults

and delay recovery from acute measles. Furthermore, $CD4^+$ T cells in the absence of $CD8^+$ T cells and neutralizing antibody are sufficient to protect Lewis rats from measles encephalitis when immunized with recombinant vaccinia virus expressing MV proteins (BRINKMANN et al. 1991).

2.2 Measles Virus CTL in Acute Infection

Cytotoxic T cells against MV-infected targets can be demonstrated in PBL from the majority of patients with acute measles infection. A number of studies suggest that MV cytotoxicity during acute infection consists of three components: (1) natural killer cells, (2) major histocompatibility complex (MHC) class I-restricted CTL, and (3) MHC class II-restricted CTL. PBL obtained from patients with acute measles can lyse MV-infected phytohemagglutinin (PHA)-stimulated blast cells. A significant component of this cytotoxicity is not MHC restricted and can be eliminated by depleting the PBL of FC-bearing cells, suggesting that NK cells represent a major component of cytotoxicity during acute measles. Class I-restricted MV-specific CTL could still be demonstrated in PBL after depletion of FC-bearing cells, though in small magnitude (KRETH et al. 1979; SISSONS et al. 1985; VAN BINNENDIJK et al. 1990).

Recently, we have examined the MV CTL response in PBL from 18 patients with acute measles 1 week–2 months after rash. Cytotoxicity against uninfected and MV-infected autologous EBV-transformed B cell targets was measured in unstimulated PBL and in PBL stimulated with MV in vitro. While MV CTL was low in unstimulated PBL ($<5\%$ specific lysis), 15 of the 18 patients had significant elevation in MV CTL when MV-stimulated PBL were used as effectors, with a mean specific lysis of $40 \pm 14.4\%$ (Fig. 2). The CTL response was MV specific, as MV-stimulated PBL produced much lower lysis of influenza virus-infected targets, as shown in Table 1. Analysis of the MHC

Table 1. Virus specificity of the measles virus (MV) cytotoxic T lymphocyte (CTL) response during acute measles shown as percent lysis of autologous virus-infected Epstein-Barr virus (EBV)-transformed B cell targets at effector-to-target ratio of 40:1

Case	Effectors	Targets		
		Measles-infected	Infuenza-infected	Uninfected
1	Unstimulated	1	10	2
	Measles	60	19	17
	Influenza	4	61	2
2	Unstimulated	12	5	1
	Measles	59	12	2
	Influenza	8	12	3
3	Unstimulated	3	3	0
	Measles	26	5	1
	Influenza	25	38	3

restriction of the CTL response from three patients with acute infection against class I- or class II-matched targets suggested that both class I- and class II-restricted CTL exist in the PBL from those patients when tested within 3 weeks after infection (E. Hurwitz, S. Dhib-Jalbut, and H. F. Mc Farland, unpublished data). In a recent study, stimulation of PBL from children 3–4 weeks after contracting acute measles, with MV-infected EBV-transformed B cells as antigen-presenting cells resulted in the generation of MV-specific clones. These clones were predominantly $CD8^+$ class I-restricted CTL, though $CD4^+$ class II-restricted CTL were also generated, but in smaller frequency (VAN BINNENDIJK et al. 1989). Collectively, these studies indicate that both class I-restricted and class II-restricted MV-specific CTL are generated during acute measles. While cell-mediated immunity is important for recovery from acute measles infection, the relative significance of class I- and class II-restricted CTL in recovery from the acute infection and in establishing lifelong immunity to MV is not understood. Three of the 18 patients with acute measles that we studied had low MV CTL at two time points following rash. These three patients recovered without complications, suggesting that MV CTL in low frequency may be sufficient for recovery or that MV CTL is not critical for recovery.

2.3 Measles Virus CTL in Post-Measles Encephalomyelitis

PME is a serious complication of measles that occurs within 3 weeks after the infection and has a 10%–20% fatality rate. Pathologically, the CNS shows a striking resemblance to post-rabies vaccine encephalomyelitis and experimental allergic encephalomyelitis (EAE) in animals, thus suggesting a T cell-mediated attack against the CNS. MV proteins and RNA are usually undetectable in the CNS and there is little evidence for intrathecal anti-measles antibody synthesis (JOHNSON et al. 1984). One hypothesis for the pathogenesis of PME is that MV invades the CNS and induces an autoimmune T cell response directed against a component of the CNS. This is supported by the presence of T cell reactivity to myelin basic protein (MBP) in the peripheral blood of patients with PME (JOHNSON et al. 1984) and the development of MBP-reactive T cells in the course of measles encephalitis in the rat. These MBP-reactive T cells can transfer a disease similar to EAE into naive syngeneic animals (LIEBERT et al. 1987), suggesting that PME might be a T cell-mediated disease.

A second hypothesis is that MV-specific CTL are generated in the course of PME and attack virus-infected CNS cells. In fact, FLEISCHER and KRETH (1983) have demonstrated the presence of CTL clones in the CSF of a patient with PME. More than half of the clones showed MV-specific cytotoxicity and appeared to be HLA class I-restricted (KRETH et al. 1979). Thus, it can be postulated that CTL could recognize MV antigen or a cross-reacting antigen on cerebral endothelial cells and other CNS cells, thus leading to tissue damage. However, the lack of demonstrable MV antigen in the CNS and lack of evidence for T cell cross-reactivity between MV and MBP weaken, but do not exclude, this hypothesis (RICHERT et al. 1988).

2.4 Measles Virus CTL in Subacute Sclerosing Panencephalitis

SSPE is a rare complication of measles which is characterized by a slowly progressive, persistent CNS infection with MV that eventually results in death. The neurons and oligodendrocytes contain MV genome with partially restricted MV protein synthesis, associated with parenchymatous inflammation, gliosis, and variable degrees of demyelination. The serum and CSF contain highly elevated levels of measles antibodies. The pathogenesis of SSPE is not well understood, but a number of hypotheses exist, including (a) defective virus, (b) mutation affecting the M protein, which restricts packaging and budding of the virus, (c) antibody modulation, and (d) infection with MV at a critical differentiation and maturational stage of the CNS and immune system. Another aspect of the pathogenesis of SSPE addresses the failure of the immune system to contain and eliminate the infection. In this regard, a number of hypotheses are plausible: (a) since neurons do not express MHC molecules, this could favor viral persistence, as the infected cells cannot be recognized by CTL, (b) mutations affecting viral epitopes that are critical for CTL recognition of infected cells, or (c) defective MV-specific CTL response (DHIB-JALBUT et al. 1989a).

We have examined the in vitro generation of MV CTL in PBL from patients with SSPE. PBL stimulated with the Edmonston strain of MV for 7 days were tested for lysis of MV-infected autologous EBV-transformed B cells. Since B cells express both MHC class I and class II molecules, this assay measures both MHC class I- and class II-restricted CTL. Two of the four patients had markedly reduced MV CTL compared to healthy donors. One had a moderately low MV CTL response and the fourth had a response comparable to the healthy controls. The response from the later two patients was surprisingly low in the face of an overwhelming persistent infection. Further analysis of the CTL response indicated that the bulk of the killing was mediated by NK cells. Thus, it appears that patients with SSPE may have a defect in the generation of HLA-restricted MV-specific CTL, since the CTL responses to influenza virus and MPS were intact (Fig. 3; DHIB-JALBUT et al. 1989a). As virus-specific CTL plays a role in the recovery from virus infection, a defective MV CTL response in SSPE could contribute to the lack of MV clearance from the CNS. In fact, experimental MV infection of Lewis rat brain supports a role for $CD4^+$ in protection and for $CD8^+$ T cells in recovery. Immunization of Lewis rats with vaccinia recombinants expressing the HA, F, or NC protein protected rats from MV encephalitis. This protection was not affected by $CD8^+$ T Cell depletion and appeared to be CD4 dependent

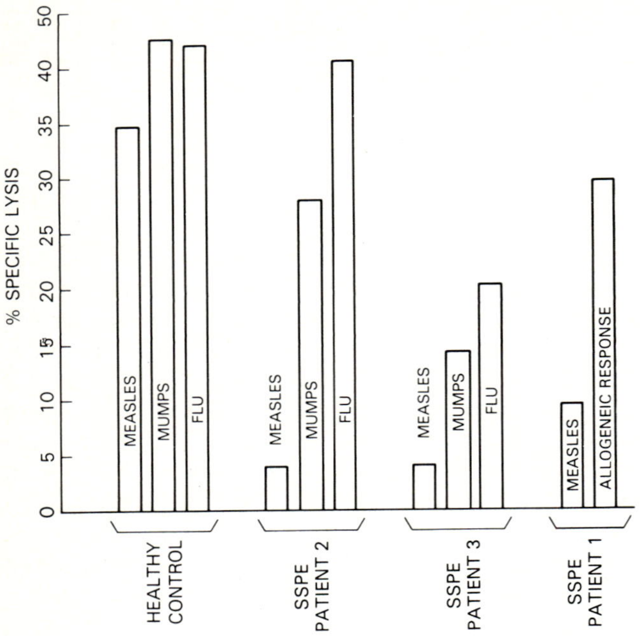

Fig. 3. Virus-specific lysis of autologous targets by measles virus (MV), mumps virus, influenza virus (*flu*), or allogeneic effectors at an effector-to-target ratio of 40: 1 in one healthy control and three subacute sclerosing panencephalitis (*SSPE*) patients. Note the specific reduction in MV cytotoxic T lymphocytes compared to other responses. (Reproduced with permission from DHIB-JALBUT et al. 1989a)

(BRINKMANN et al. 1991). However, in a different study, depletion of $CD8^+$ T cells resulted in prolonged recovery from measles encephalitis and persistence of MV antigen in neurons for a longer period (MAEHLEM et al. 1989).

2.5 Measles Virus CTL in Multiple Sclerosis

MS is an inflammatory demyelinating disease of the CNS of unknown etiology. A heightened antibody response to MV in the serum and CSF has been confirmed in a number of laboratories, but the mechanism for this increased response and its relationship to the pathogenesis of MS remain unclear (DHIB-JALBUT and MCFARLIN 1990). More recently, a defect in the generation of MHC class II-restricted MV-specific CTL has been found in PBL in the majority of MS patients, but not in healthy controls or patients with other neurologic disease of nonviral etiology (JACOBSON et al. 1985). This defect was not predominantly directed against a single MV polypeptide (Fig. 4) and

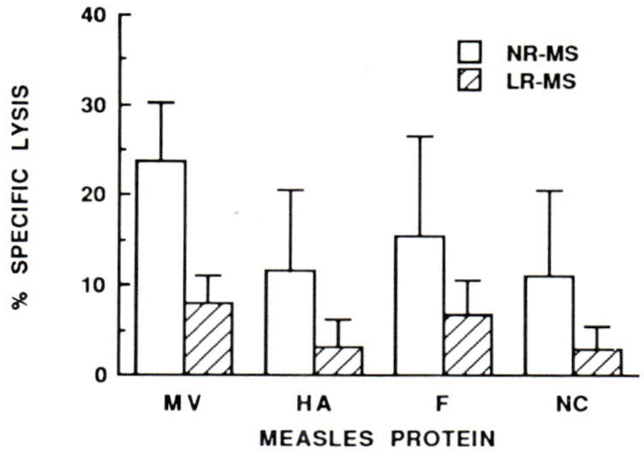

Fig. 4. Cytotoxic T lymphocyte (CTL) responses to measles virus (MV) and its polypeptides hemagglutinin (*HA*), fusion protein (F), and nucleocapsid (*NC*) in six multiple sclerosis (MS) patients with normal MV-CTL response (*NR-MS*) and eight MS patients with low MV CTL (*LR-MS*). Note that the reduction in MV CTL is not due to a defect predominantly involving a single MV polypeptide

did not appear to be due to sequestration of MV CTL in the CNS (DHIB-JALBUT et al. 1989b). The majority of MS patients with reduced MV CTL were subsequently shown to have reduced herpes simplex type I CTL, which is predominantly class II restricted (DESILVA et al. 1991). CTL responses to influenza virus and MPS, which are primarily class I restricted, were intact (JACOBSON et al. 1985; GOODMAN et al. 1989). The defect in generation of an MV CTL response is not MS specific, since it has been found in patients with SSPE (DHIB-JALBUT et al. 1989a) and in patients with human T cell leukemia/lymphoma virus type 1 (HTLV-I)-associated myelopathy (HAM/TSP; JACOBSON et al. 1990). Thus, it appears that the reduction of MHC class II-restricted CTL may be associated with viruses that establish persistence or alternatively may reflect a generalized defect in HLA class II-restricted CTL in MS. Whether the reduced MV CTL in MS is due to reduction in T-helper cell function or differentiation into effector CTL remains unclear.

3 The CTL Response to Mumps Virus

MPS is primarily a childhood infection of the salivary glands causing parotitis, and in rare cases the virus spreads to the kidneys, testicles, and CNS (WOLINSKY and SERVER 1985). Most cases of MPS complications are self-limiting and resolve without sequelae. Patients are left with lifelong immunity and antibodies are detectable in the serum of healthy individuals for many years after the infection. Like MV, MPS is a lymphotropic virus, and lymphoproliferative and virus-specific CTL responses can be generated in vitro from healthy individuals (ENSSLE et al. 1987; KRESS and KRETH 1982; GOODMAN et al. 1989). A cellular immune response to MPS probably develops early during infection and inflammation is readily present in affected tissue. This is further supported by the demonstration of CTL during acute infection and in the spinal fluid of patients with MPS meningitis (KRETH et al. 1982; FLEISHER and KRETH 1983).

The precursor frequency of MPS-specific CTL in PBL of healthy individuals with previous exposure of MPS has been estimated to range between 1/500 and 1/8000 (ENSSLE et al. 1987). Studies of the phenotype and HLA restriction of MPS-specific CTL in peripheral blood and CSF of patients with MPS meningitis demonstrated that MPS CTL are largely $CD8^+$ and HLA class I restricted (KRESS and KRETH 1982; ENSSLE et al. 1987). Interestingly, the magnitude of the CTL response to MPS appears to be linked to certain HLA class I antigens, specifically HLA B7, B13, B27, and Bw52, which may in turn influence the clinical expression of MPS infection.

Recently, GOODMAN et al. (1989) demonstrated that PBL from healthy individuals cultured with MPS for 7 days generated MPS CTL that contained both $CD4^+$ class II-restricted and $CD8^+$ class I-restricted CTL, when EBV-transformed class I and class II expressing B cells were used as targets. The magnitude of this MPS-specific CTL response in healthy individuals ranged between 20% and 65% with a mean of 44.5%. Approximately 40%–60% of the MPS CTL response appeared to be $CD4^+$ class II restricted.

The significance of MPS CTL in recovery from MPS infection and maintenance of long-term immunity is not clear. It appears that MPS CTL are not that critical for recovery from acute infection, as the course of mumps in the hamster is not altered by cyclophosphamide treatment and persistent MPS infection has not developed in immunocompromised patients (WOLINSKY et al. 1985).

4 The CTL Response to Respiratory Syncytial Virus

RSV was initially classified among the paramyxoviruses based on its morphological features. However, because the helical NC of RSV was found to be smaller than that of other paramyxoviruses, it is currently classified in a separate genus, the pneumoviruses. RSV consists of six major structural proteins: a nucleoprotein (N), surface protein (SP), fusion protein (F), matrix protein (M), glycoprotein (G), and phosphoprotein (P; MCINTOSH and CHANOCK 1985).

RSV is a common cause of respiratory tract infection and pneumonia in infants and children. Children lacking cellular immunity are unable to clear the infection effectively and continue to shed virus for several months. In contrast to other paramyxoviruses, immunity to RSV is not permanent and reinfection is common. While serum antibodies to RSV are believed to play a role in protection, cell-mediated immunity is considered important for recovery (MCINTOSH and CHANOCK 1985).

Cell-mediated cytotoxicity has been examined in infants following naturally acquired RSV infection. PBL from these infants showed a peak cytotoxicity against autologous RSV-infected mononuclear cells at 1 week postinfection. Cytotoxicity against allogeneic targets was also present but lower in magnitude, suggesting that RCV cytotoxicity is mediated at least in part by virus-specific CTL (CHIBA et al. 1989).

In healthy adults, RSV-specific memory CTL can be stimulated in vitro by culturing PBL with virus. These PBL manifest class I-restricted CTL response against B lymphoblastoid cell targets infected with recombinant vaccinia viruses expressing the different RSV individual proteins. While N, M, SH, F, and a 22 K were recognized by RSV CTL, most of the response was directed against the N. Fewer individuals developed CTL to G or P (CHERRIE et al. 1992). Similar studies in the mouse identified 22 K protein as the major target for CTL followed by the F and N, but little recognition of P and no recognition of G, SH, or M was observed (NICHOLAS et al. 1990). Interestingly, repeated stimulation of murine CTL with RSV resulted in selective growth of $CD4^+$ CTL which recognized virus-infected but not vaccinia recombinant-infected targets. This is consistent with the concept that different antigen-processing pathways are used by $CD4^+$ and $CD8^+$ CTL. The same group of investigators examined RSV CTL in healthy adults. CTL could be generated after two repeated stimulations of PBL with RSV, and as with MV the CTL was primarily $CD4^+$class II restricted (NICHOLAS et al. 1990).

Collectively, these studies seem to indicate that RSV-specific $CD8^+$ CTL develops during acute infection and probably plays a role in recovery. $CD4^+$ CTL, on the other hand, which can be generated from healthy adults, may play a role in protection. It would be significant to determine whether $CD4^+$ CTL can be generated from infants and children after recovery from RSV infection. If not, this would suggest a role for $CD4^+$ CTL in protection, as infants and children are at higher risk than adults for reinfection.

5 Conclusions

The CTL response to the three human paramyxoviruses, MV, MPS, and RSV, share a number of characteristics. All three viruses can generate both $CD8^+$ class I-restricted and $CD4^+$ class II-restricted CTL. The $CD8^+$ response seems to play a role in the pathogenesis of and recovery from acute infection. The $CD4^+$ response is demonstrable after recovery and may play a role in long-term immunity. With MV and RSV, a CTL response to the individual viral polypeptides has been demonstrated, and further analysis of the fine specificity of CTL epitopes should provide useful information for the development of effective subunit vaccines.

Acknowledgement. We would like to thank Ms. Berñadette Pasko for preparing the typescript for the manuscript.

References

Brinckmann UG, Bankamp B, Reich A, Ter Meulen V, Liebert UG (1991) Efficacy of individual measles virus structural proteins in the protection of rats from measles encephalitis. J Gen Virol 72: 2491–2500

Cherrie AH, Anderson K, Wertz GW, Openshaw PJ (1992) Human cytotoxic T cells stimulated by antigen in dendritic cells recognize the N, SH, F, M, 22 K, and 1 b proteins of respiratory syncytial virus. J Virol 66: 2102–2110

Chiba Y, Higashidate Y, Suga K, Honjo K, Tsutsumi H, Ogra PL (1989) Development of cell-mediated cytotoxic immunity to respiratory syncytial virus in human infants following naturally acquired infection. J Med Virol 28: 133–139

DeSilva SM, McFarland HF (1991) Multiple sclerosis patients have reduced HLA class II-restricted cytotoxic responses specific for both measles and herpes virus. J Neuroimmunol 35: 219–226

Dhib-Jalbut S, McFarlin DE (1990) Immunology of multiple sclerosis. Ann Allergy 64: 433–444

Dhib-Jalbut S, Jacobson S, McFarlin DE, McFarland HF (1989a) Impaired human leukocyte antigen-restricted measles virus-specific cytotoxic T-cell response in subacute sclerosing panencephalitis. Ann Neurol 25: 272–280

Dhib-Jalbut S, McFarlin DE, McFarland HF (1989b) Measles virus polypeptide specificity of the cytotoxic T-lymphocyte response in multiple sclerosis. J Neuroimmunol 21: 205–212

Enssle KH, Wagner H, Fleischer B (1987) Human mumps virus-specific cytotoxic T lymphocytes. Quantitative analysis of HLA restriction. Hum Immunol 18: 135–149

Fleischer B, Kreth HW (1983) Clonal expansion and functional analysis of virus-specific T-lymphocytes from cerebrospinal fluid in measles encephalitis. Hum Immunol 7: 239–248

Fleisher B, Kreth HW (1983) Clonal analysis of HLA-restricted virus-specific cytotoxic T-lymphocytes from cerebrospinal fluid in mumps meningitis. J Immunol 150: 2187

Goodman AD, Jacobson S, McFarland HF (1989) Virus-specific cytotoxic T lymphocytes in multiple sclerosis: a normal mumps virus response adds support for a distinct improvement in the measles virus response. J Neuroimmunol 22: 201–209

Jacobson S, Richert JR, Biddison WE, Satinsky A, Hartzman RJ, McFarland HF (1984) Measles virus-specific $T4^{+}$ human cytotoxic cell clones are restricted by class II HLA antigens. J Immunol 133: 754–757

Jacobson S, Flerlage ML, McFarland HF (1985) Impaired measles virus-specific cytotoxic T cell responses in multiple sclerosis. J Exp Med 162: 839–850

Jacobson S, Rose JW, Flerlage ML, Mingioli EE, McFarlin DE, McFarland HF (1987) Measles virus-specific human cytotoxic T-cells generated in bulk culture. Analysis of measles virus antigenic specificity. In: Mahy B, Kolakofsky D (eds) The biology of negative strand virus, chap 37. Elsevier Science, New York, pp 283–289

Jacobson S, Gupta A, Matson D, Mingioli E, McFarlin DE (1990) Immunological studies in tropic spastic paraparesis. Anal Neurol 27: 149–156

Johnson RT, Griffin DE, Hirsch RL, Wolinsky JS, Roedenbeck S, Lindo de Suriano I, Vaisberg A (1984) Measles encephalomyelitis—clinical and immunologic studies. N Engl J Med 330: 137–145

Kaplan LJ, Daum RS, Smaron M, McCarthy LA (1992) Severe measles in immunocompromised patients. JAMA 267: 1237–1241

Kingsbury DW (1985) Orthomyxo- and paramyxoviruses and their replications. In: Fields BN et al. (eds) Virology, chap 50. Raven, New York, pp 1157–1178

Kress HG, Kreth HW (1982) HLA-restriction of secondary mumps—specific cytotoxic T lymphocytes. J Immunol 129: 844

Kreth HW, Ter Meulen V, Eckert G (1979) Demonstration of HLA restricted killer cells in patients with acute measles. Med Microbiol Immunol 165: 203–214

Kreth HW, Kress L, Kress HG, Oh HF, Eckert G (1982) Demonstration of primary cytotoxic T-cells in venous blood and cerebrospinal fluid of children with mumps meningitis. J Immunol 128: 2411

Liebert UG, Linington C, ter Meulen V (1987) Induction of autoimmune reactions to myelin basic protein in measles virus encephalomyelitis in Lewis rats. J Neuroimmunol 17: 103

Lucas CJ, Biddison WE, Nelson DL, Shaw S (1983) Killing of measles virus-infected cells by human cytotoxic T-cells. Infect Immunol 38: 226–232

Maehem J, Olsson T, Love A, Klareskog L, Norrby E, Kristensson K (1989) Persistence of measles virus in rat brain neurons is promoted by depletion of CD8$^+$ T-cells. J Neuroimmunol 21:149–155
McIntosh K, Chanock RM (1985) Respiratory syncytial virus. In: Fields BN et al. (eds) Virology, chap 54. Raven, New York, pp 1285–1304
Nicholas JA, Rubino KL, Levely ME, Adams EG, Collins PL (1990) Cytolytic T-lymphocyte responses to respiratory syncytial virus: effector cell phenotype and target proteins. J Virol 64:4232–4241
Norrby E (1985) Measles. In: Fields BN et al. (eds) Virology, chap 55. Raven, New York, pp 1305–1321
Richert JR, Robinson ED, Reuben-Bornsidl CA, Johnson AH, McFarland HF, McFarlin DE, Hartzman RT (1988) Measles virus-specific T-cell clones. Studies of alloreactivity and antigenic cross reactivity. J Neuroimmunol 19:59–68
Sha BE, Harris AA, Benson CA, Atkinson WL, Urbanski PA, Stewart JA, Williams WW, Murphy RL, Phair JP, Levin SA, Kessler HA (1991) Prevalence of measles antibodies in asymptomatic human immunodeficiency in virus-infected adults. J Infect Dis 164:473–475
Sissons Patrick JG, Colby SD, Harrison WO, Oldstone MB (1985) Cytotoxic lymphocytes generated in vivo with acute measles virus infection. Clin Immunol Immunopathol 34:60–68
Suringa DWR, Bank LJ, Ackerman AB (1970) Role of measles virus in skin lesions and Koplik's spots. N Engl J Med 283:1139–1142
Van Binnendijk RS, Poelen MCM, DeVries P, Voorma HO, Osterhaus ADME, Uytdehaag FGCM (1989) Measles virus-specific human T cell clones. Characterization of specificity and function of CD4$^+$ helper/cytotoxic and CD8$^+$ cytotoxic T-cell clones. J Immunol 142: 2897–2854
Van Binnendijk RS, Poelen MCM, Kuijpers KC, Osterhaus ADME, Uytdehaag FGCM (1990) The predominance of CD8$^+$ T cells after infection with measles virus suggests a role for CD8$^+$ class I MHC-restricted cytotoxic T lymphocytes (CTL) in recovery from measles. Clonal analysis of human CD8$^+$ class I MHC-restricted CTL. J Immunol 144: 2394–2399
Wolinsky JS, Server AC (1985) Mumps virus. In: Fields BN et al. (eds) Virology, chap 53. Raven, New York, pp 1255–1284
Wu VH, McFarland HF, Mayo K, Hanger L, Griffin DE, Dhib-Jalbut S (1993) Measles virus-specific cellular immunity in patients with vaccine failure. J Clin Microbiol 31:118–122

Cytotoxic T Cells and Human Herpes Virus Infections

L. K. BORYSIEWICZ[1] and J. G. P. SISSONS[2]

[1] Department of Medicine, University of Wales College of Medicine, Heath Park, Cardiff CF4 4XN, UK

[2] Department of Medicine, University of Cambridge Clinical School, Hills Road, Cambridge CB2 2QQ, UK

Current Topics in Microbiology and Immunology, Vol. 189

1 Introduction

This chapter reviews what is currently known about the role of cytotoxic T lymphocytes (CTL) in human herpesvirus infections.

1.1 Importance of Human Herpesviruses

Herpesviruses are large, double-stranded DNA viruses which infect a wide range of vertebrate hosts. They all establish persistent or latent infection and are widely prevalent in their host population. Members of the family cause a range of diseases, distinct syndromes often being associated with primary infection and with reactivation of the virus from latency—some members have in addition been associated with certain tumours. The tendency of all these viruses to reactivate and produce disease in the context of immunosuppression has long suggested that the immune response—particularly the T cell response—plays an important role in controlling infection in the normal host. It is important to emphasise that although this review is restricted to CTL, they are only one component of the immune response: T cells mediating help and delayed hypersensitivity and antibody may also play a role in the control of infection.

1.2 Methods and Problems in the Study of Human Virus-Specific CTL

Although we cannot deal in any detail with the methodology of CTL assays, there are some points worth emphasising and which are important to bear in mind when interpreting published work in this area. In murine experimental models, CTL can be directly sampled from lymph nodes and spleen at any chosen time point during an experimentally induced virus infection. The contribution of CTL to the overall immune response can be analysed by depletion and adoptive transfer experiments and the role of immunogenetic factors clearly defined using inbred strains. In humans by contrast, peripheral blood is usually the only compartment that can be sampled, and it is often difficult to obtain samples during a natural virus infection. These two factors in particular mean that in order to obtain activated effector T cells, memory cells present in peripheral blood have to be subjected to secondary restimulation in vitro and, depending on the protocol adopted, this can lead to a selection bias. It is in addition obviously more difficult to examine the influence of immunogenetic factors and very difficult to do clinical experiments to establish an unequivocal protective role for CTL.

The method adopted for assaying human virus specific CTL varies, although almost all culminate in an assay of chromium release from appropriately major histocompatibility complex (MHC)-matched target cells infected with virus or expressing a particular virus gene product; the extent to which MHC restriction and natural killer cell activity are controlled varies considerably between reported studies. Secondary in vitro restimulation of CTL precursors is usually achieved by culturing peripheral blood mononuclear cells (PBMC) with virus or virus-infected cells for 1–2 weeks, with addition of exogenous interleukin-2 (IL-2); alternatively, a recombinant protein or peptide epitope may be used. Usually during this period the cells are kept as a bulk culture, but they may be set up in limiting dilution culture, which allows CTL precursors to be quantitated. The T cells may also be cloned following restimulation and before being assayed, which potentially introduces a major selection step, although it has proved particularly difficult to maintain human $CD8^+$ CTL clones. A better understanding of the factors involved in T cell activation should lead to improvement in the efficiency with which T cells can be activated in vitro: as an example, the recent recognition of the interaction between the B7 molecule on the antigen-presenting cell and CD28 on the T cell, as a major co-stimulatory signal inducing IL-2 production even in $CD8^+$ CTL, is likely to be important in this context (see SCHWARTZ 1992).

2 Virology and Biology of Herpesvirus Infections

This topic is dealt with in depth in recent reviews and texts (ROIZMAN 1990). Here we only deal briefly with aspects relevant to the immunolgy of the infection.

2.1 Human Herpesviruses

There are currently six recognized human herpesviruses, whose nomenclature is shown in Table 1.

2.2 Genome Structure

Herpesviruses have large, double-stranded DNA genomes coding for 150–200 proteins. All have now been completely sequenced at the DNA level, revealing blocks of conserved genes but with others distinctive for a particular virus (BAER et al. 1984; DAVISON and SCOTT 1986; CHEE et al. 1990). These homologies justify the grouping of vricella zoster virus (VZV) and herpes simplex virus types 1 and 2 (HSV-1 and -2) as alpha herpesviruses and show that the relatively recently isolated human herpesvirus 6 is closer to

Table 1. The human herpesviruses

Common name	Designation	Subfamily	Genome size (kilobase pairs)	Site of persistence
Herpes simplex virus 1	Human herpesvirus 1	α	152	Neurons (sensory ganglia)
Herpes simplex virus 2	Human herpesvirus 2	α	152	Neurons (sensory ganglia)
Varicella zoster virus	Human herpesvirus 3	α	125	Neurons (sensory ganglia)
Epstein-Barr virus	Human herpesvirus 4	γ	172	B cells, oropharyngeal epithelium
Cytomegalovirus	Human herpesvirus 5	β	235	Monocytes, epithelial cells
	Human herpesvirus 6	β	170	Monocytes, T cells

cytomegalovirus (CMV) than to the other herpesviruses. The sequencing has also revealed genes with homology to cellular genes: in the case of human CMV (HCMV), which has the largest genome of the human herpesviruses, its extra size is accounted for by many extra open reading frames coding for glycoproteins; some of these show homology to human cellular proteins including the class I MHC heavy chain (BECK and BARRELL 1988) and G protein-coupled receptors (CHEE et al. 1990). Homologies to cellular genes exist in other herpesviruses, for instance to the C3b receptor in HSV and to IL-10 in Epstein-Barr virus (EBV).

2.3 Regulation of Gene Expression

Herpesvirus genes are expressed in a temporally regulated sequence designated immediate–early (IE), early and late (this nomenclature is not used in the case of EBV, although there are broadly analogous genes expressed in its latent and lytic cycles). The control of transcription of these genes is critical to initiating the infection; it is also important to emphasise that gene expression may be restricted in some tissues, particularly those in which the virus persists or is latent. In addition expression of viral genes may be linked to the state of activation or differentiation of the cell; for instance, HCMV is not expressed in monocytes, but is when these cells differentiate into tissue macrophages (IBANEZ et al. 1991; TAYLOR-WIEDEMAN et al. 1991), and the extent of EBV gene expression varies in different B cell and Burkitt lymphoma lines (MURRAY 1992; ROWE et al. 1992). These differences in viral gene expression dependent on cell type probably result from the presence or absence of specific cellular transcription factors which bind to the promoter regulatory region of viral genes. Such cellular factors are particularly important in controlling transcription of IE genes (GARCIA-BLANCO and CULLEN 1991). Viral proteins may have transactivating functions; for instance, the HSV vMW65 virion protein transactivates the IE genes of the virus (ROIZMAN and SEARS 1990), and the IE proteins themselves transactivate expression of the early viral genes and may autoregulate their own expression (STAMMINGER and FLECKENSTEIN 1990). In addition, certain viral IE genes may also be capable of transactivating cellular genes (HAGEMEIER et al. 1992). These considerations are all important in determining which gene products are available for recognition by the immune response.

2.4 Latency and Persistence

The alpha herpesviruses establish latent infection in neuronal cells in sensory ganglia, whence they periodically reactivate and are transported down the axon to infect cells at the nerve ending. There is little or no viral transcription occurring for most of the time, the only consistent transcript detected in HSV latency in neurons being the latency-associated transcript (LAT), which is not expressed as protein and whose function is uncertain (see ROIZMAN and SEARS 1990). Whether the infected neuron is susceptible to attack by CTL or other immunological mechanisms is again uncertain, but seems unlikely.

EBV persists in a small fraction of B cells. In lymphoblastoid B cell lines in vitro, the EBV DNA exists as an episome which is maintained at constant copy number as the cells divide; EBV also persists in epithelial cells in the oropharynx. There is probably restricted EBV gene expression in the B cell, whereas viral replication occurs in the oropharyngeal epithelium. The extent to which these two sites of persistence are accessible to the immune response in vivo is uncertain, although it is assumed that the potential of EBV to induce B cell transformation is somehow restrained in vivo by the T cell response (ROWE et al. 1992).

There is less certainty concerning the sites at which HCMV persists, but HCMV DNA is present in monocytes in normal virus carriers and probably also persists at epithelial sites such as the salivary gland. There appears to be little or no transcription in monocytes, and the extent to which either site of persistence is accessible to the immune response is again unknown (TAYLOR-WIEDEMAN et al. 1991; IBANEZ et al. 1991). Human herpesvirus 6 replicates in activated T cells in vitro and is detectable in PBMC, probably in T cells and monocytes in vivo (see THOMSON et al. 1991).

The mechanisms by which latency is maintained are in general relatively poorly understood, but are a major area of current study in herpes virology.

3 CTL in Experimental Herpes Virus Infections

Although this review deals with human CTL, there is a great deal of work on experimental herpesvirus infections in animals, especially the mouse. This is reviewed elsewhere (MESTER and ROUSE 1991; KOSZINOWSKI et al. 1990), but we refer briefly to those aspects which seem particularly pertinent to the role of CTL in human herpesvirus infections.

3.1 Herpes Simplex Virus in the Mouse

Although a number of animal models of HSV infection have been described, how representative they are of the human disease, and thus how well the recorded immune responses accord with man, is arguable (MESTER and ROUSE 1991). The mouse has been particularly used to study challenge inoculation by a number of routes—the most representative probably being the ear clearance model, particularly when used to study local protection. The zosteriform spread model has been used to examine factors that may be important in recrudescent disease, although no reproducible model of latency and reactivation is established in the mouse (MESTER and ROUSE 1991). $CD4^+$ delayed-type hypersensitivity (DTH) cells were of particular importance in the clearance of low-dose virus such as with local ear inoculation, whereas $CD8^+$ CTL protect against high-dose infection. Murine $CD8^+$ CTL clear HSV-infected cells in vivo in immunodepletion studies (NASH et al. 1987). In addition, shedding of virus at the site of infection may be reduced by $CD8^+$ T cells (BONNEAU and JENNINGS 1988). These murine $CD8^+$ CTL recognise both viral glycoproteins, and in limiting dilution studies approximately 30% of the CTL response is directed against IE antigens (MARTIN et al. 1988). In order to study recurrent disease, the guinea pig vaginal model has been used: genital infection can be produced and reactivation induced by similar mechanisms to those which reactivate human infection. For these reasons it is used as a model in which to test candidate HSV vaccines. The obvious disadvantage is that there is less scope for manipulating the immune system: the absence of specific T cell markers in this animal makes detailed examination of the CTL response difficult, although CTL have been described (STANBERRY 1991).

3.2 Murine Cytomegalovirus Infection

The cytomegaloviruses are species specific and HCMV does not infect the mouse. However, murine CMV (MCMV), although not able to productively infect human cells, has somewhat similar biology to HCMV and has been used as a model of HCMV infection.

Koszinowski, Reddehase and colleagues have reported extensive studies of the role of CTL in MCMV infection using the Balb/c strain of mice and have recently reviewed these in detail (KOSZINOWSKI et al. 1990). Following infection via the footpad, MCMV-specific CTL are recoverable from the draining lymph node. In acute infection there is a dominant population (about 50% of the MCMV-specific $CD8^+$ CTL) detectable showing specificity for determinants expressed at IE times (principally the major IE protein of MCMV). Other CTL are specific for unidentified structural proteins of the virion, detected by their ability to lyse target cells infected with UV-inactivated virus (REDDEHASE et al. 1984); during persistent infection, the former population predominate. In adoptive transfer experiments into irradiated MCMV-infected mice, $CD8^+$ CTL specific for the 89-kDa IE protein transfer protection from lethal MCMV infection (REDDEHASE et al. 1987). The clearance of MCMV from the lungs of infected mice has also been shown to be mediated by the same $CD8^+$ population (M. Reddehase, personal communication). This protective CTL response has been mapped to a pentapeptide from the IE protein which binds to $H2\text{-}L^d$. Mice vaccinated with a recombinant vaccinia virus expressing the 89-kDa protein are also protected from subsequent lethal challenge (JONJIC et al. 1988). In CD8-depleted mice, $CD4^+$ cells can transfer a degree of protection from lethal challenge, but require an accessory cell for the effect. However, the clearance of MCMV from the salivary gland has been shown to depend completely on $CD4^+$ cells and to be mediated by γ-interferon produced by these cells (JONJIC et al. 1989; LUCIN et al. 1992).

Whilst this elegant work clearly demonstrates a role for IE-specific CTL in this model, it has to be remembered that the work does all apply to a single inbred mouse strain, and the extent to which the results are generalisable to HCMV infection remains to be determined.

It is worth mentioning that MCMV is also one of the few experimental virus infections in which a protective role for NK cells has been convincingly shown (BUKOWSKI et al. 1985).

3.3 Other Experimental Infections

A recently described herpesvirus isolated from wild mice seems to have the characteristics of a murine equivalent of EBV (SUNIL-CHANDRA et al. 1992a, b). If initial studies are borne out, this could become an extremely valuable model, as the only other experimental model of EBV is EBV infection of cottontop tamarinds. This latter model has been used to study candidate EBV vaccines, particularly that based on gp340; studies to date have provided evidence of protection against EBV-induced lymphomagenesis in this model (MORGAN 1991).

4 CTL in Herpes Simplex Virus Infection

4.1 Asymptomatic Persistent Infection

As described above, HSV-specific CTL responses have been extensively investigated in a variety of murine models. In these experimental systems, such cells have been shown to mediate a variety of protective responses, but a major difficulty in their interpretation is knowing how representative they are of human HSV infection, particularly in considering episodes of reactivation in man (MESTER and ROUSE 1991). Initial attempts to demonstrate the presence of CTL activity against HSV in man frequently resulted in generation of $CD4^+$ CTL in vitro (TORPEY and LINDSLEY 1989; MESTER and ROUSE 1991; SCHMID 1988) at a relatively high precursor frequency of 1/4000–1/20 000 PBMC. These experiments predominantly used a secondary in vitro stimulation protocol where UV-inactivated virus was added to PBMC. It may therefore not be surprising that as with other human herpesviruses when this protocol is used, e.g. HCMV (BORYSIEWICZ et al. 1983), T cells with this phenotype predominated. ZARLING et al. (1986, 1988; YASUKAWA and ZARLING

1985) investigated the specificity of this CD4 CTL response in a series of human $CD4^+$ CTL clones. They showed that such isolated clones recognised HSV glycoproteins gB and gD (ZARLING et al. 1986). In addition, these $CD4^+$ CTL clones secreted IL-2 (ZARLING et al. 1988).

However, YASUKAWA et al. (1989) found that by altering the in vitro stimulation protocol to co-culture PBMC with HSV-infected cells, they were able to establish $CD8^+$ CTL lines. These studies have been extended by TIGGES et al. (1992), who used HSV-2-infected irradiated stimulator cells with PBMC from a subject with recurrent genital herpes. They then cloned the responding cells by limiting dilution, using phytohemagglutinin (PHA) and allogeneic irradiated feeder cells, to generate seven HSV-specific $CD8^+$ CTL clones. Six of these were HSV-2 specific, one cross-reacting with HSV-1. One clone recognised IE viral antigens, whereas five recognised virion proteins that did not require de novo viral protein synthesis in the infected target cells. Only one of the clones was further defined using HSV-2 vaccinia recombinants as being gD-2 specific.

4.2 Disease and Immunosuppression

The cutaneous location of HSV-1 lesions makes them ideally suited to examine infiltration by T cells. CUNNINGHAM et al. (1985) found that $CD4^+$ T cells are the first cells to infiltrate a cutaneous lesion, together with macrophages, although whether these cells mediate direct cytotoxicity or are the counterpart of the murine $CD4^+$ DTH cells is unknown. By day 3, $CD8^+$ T cells are present in the lesion. Similarly, in the cornea affected by herpetic keratitis, both $CD4^+$ and $CD8^+$ T cells were present, together with increased MHC class II expression on surrounding epithelium (CUNNINGHAM and NOBLE 1989).

5 CTL in Varicella Zoster Virus Infection

There have been a relatively limited number of studies of VZV-specific CTL, and there is no specific animal model. However, one interesting clinical aspect of VZV infection is that a live vaccine has been used for some time and offers a chance to assess the induction of effector T cells in vivo.

5.1 Asymptomatic Persistent Infection

VZV-specific CTL can be generated from the peripheral blood of most normal seropositive patients. The generation of both $CD4^+$ and $CD8^+$ CTL has been reported, with more concentration on $CD4^+$ CTL in published work. It seems probable that the phenotype of the responding cell to a considerable extent reflects the method of secondary in vitro stimulation: most studies have used inactivated VZV antigen, whereas few have used VZV-infected cells. The target cells used are usually VZV-infected lymphoblastoid cell lines, but expression of VZV may not be complete in such cells. Infected fibroblasts have also been used. VZV is characteristically strongly cell associated, which leads to difficulty in achieving synchronous infection of target cells. It is presumably this difficulty in using a system which allows endogenous expression of VZV that has led to the use of inactivated VZV antigen for in vitro restimulation.

HAYWARD et al. (1986a) first reported the generation of $CD4^+$ VZV-specific CTL clones. Arvin and colleagues (see ARVIN 1992) have also reported the generation of VZV-specific CTL using inactivated VZV antigen. They also reported that if $CD4^+$ cells were removed from these experiments at the beginning of the period of in vitro restimulation, then $CD8^+$ CTL could be generated (ARVIN et al. 1991). This lack of requirement for $CD4^+$ cells to generate a $CD8^+$ response and the ability of inactivated antigen to generate a CD4 or CD8 CTL response with equal facility are both noteable. HICKLING et al. (1987) used VZV-infected autologous B cell lines to restimulate, and fibroblasts (which do not express class II MHC molecules) as the target cells, and found that class I-restricted CTL could be demonstrated in four subjects. The frequency of VZV-specific T cells generated by this latter protocol was also estimated by limiting dilution analysis (LDA): the reported frequencies varied from 11 to 30 per 10^6 T cells (HICKLING et al. 1987). Direct VZV-specific cytotoxicity for infected fibroblasts in unstimulated PBMC had been detected by BOWDEN et al. (1985) and reported to be partially HLA restricted; however, such direct class I-restricted cytotoxicity was not detectable by HICKLING et al. (and is in general unusual in normal individuals in any persistent infection). Particular care must be taken to exclude induced NK cytotoxicity in this system (HAYWARD et al. 1986b).

It has been suggested that the frequency of VZV-specific CTL is maintained by subclinical reactivation. There are reports of VZV DNA being detected in the PBMC of immunosuppressed individuals after bone marrow transplantation and of this correlating with return of proliferative T cell responses (WILSON et al. 1992). Whether VZV DNA is intermittently detectable in normal individuals is not reported.

5.2 Disease and Immunosuppression

In one report of studies on subjects with varicella, direct CTL cytotoxicity could not be demonstrated using fresh PBMC, although it was present after restimulation as in normal VZV carriers (DIAZ et al. 1989).

There are several reports of subjects who have been studied after an episode of herpes zoster. The frequency of T cells responding to VZV antigen in proliferation assays is increased for up to 2 years following zoster, compared to age-matched controls (HAYWARD et al. 1991). Conversely, the frequency of proliferating T cells declines with increasing age in subjects who have *not* had clinical zoster, as assessed by LDA and crude proliferation assays (HAYWARD and HERBERGER 1987).

The frequency of $CD8^+$ class I-restricted VZV-specific CTL assessed by LDA has also been reported to be increased following zoster in a few subjects (HICKLING et al. 1987).

5.3 Specificity for Viral Gene Products

Using recombinant vaccinia viruses expressing the IE62 protein and glycoprotein I of VZV, precursor frequencies of $CD4^+$ and $CD8^+$ CTL specific for these two proteins, assessed by LDA, have been reported to be similar in normal virus carriers—of the order of 1/100 000, although the range was very wide (ARVIN et al. 1991). The same authors also found that CTL with the same specificity could be induced by vaccination of VZV-uninfected volunteers with the Oka VZV vaccine (SHARP et al. 1992). CTL specific for gpIV and gpV of VZV have also been reported with similar frequencies to those cited above (SHARP et al. 1992). There are no reports of specificity for other VZV proteins, the above appearing to be the only ones reported in the literature.

6 CTL in Epstein-Barr Virus Infection

Unlike the generation of most human virus-specific CTL, which requires secondary in vitro stimulation with autologous virus-infected cells, EBV CTL activity was identified by MOSS et al. (1978) after observing regression of EBV-transformed B cells in vitro. In vitro infection of PBMC from seronegative subjects with EBV resulted in the outgrowth of EBV nuclear antigen (EBNA)-positive B cell lines; however, in seropositive subjects, initial outgrowth of colonies was halted. This regression was T cell dependent and could be diluted out such that a quantitative estimate of the relative number of T cells mediating this effect could be made. Further studies allowed the establishment of classical $CD8^+$, MHC class I-restricted CTL lines and clones (RICKINSON et al. 1981; MOSS et al. 1988). Although most further studies have focused on the nature, specificity and role of these $CD8^+$ cells, as with the other herpesviruses, $CD4^+$, MHC class II-restricted CTL have also been described (MOSS et al. 1988).

6.1 Asymptomatic Persistent Infection, Disease and Immunosuppression

The role of EBV-specific cytotoxic T cells in protecting the host against the range of clinical disease associated with the virus remains an important issue, but one that is difficult to approach directly, particularly in the absence of a good animal model of disease. No specific $CD8^+$ CTL activity can be found in seronegative subjects, but shortly after symptomatic primary infection with infectious mononucleosis CTL are present in PBMC (ENSSLE and FLEISCHER 1990), and this response persists at a high precursor frequency throughout life (RICKINSON et al. 1981; ENSSLE and FLEISCHER 1990; ALP et al. 1990).

Clinically, EBV has been associated with a number of diseases which are particularly apparent in immunosuppressed subjects, namely monoclonal and polyclonal B cell lymphoma (NALESNIK 1991), oral hairy leukoplakia and interstitial pneumonitis. In addition, Burkitt's lymphoma and nasopharyngeal carcinoma are strongly associated with EBV, although association with immunosuppression in this context is less well established. Indeed, the tumour cells of Burkitt's lymphoma may have a phenotype which expresses only EBNA-1, together with reduced expression of adhesin molecules, thereby precluding lysis by CTL (GREGORY et al. 1988). Following immunosuppression with a range of pharmacological agents, there is increased shedding of EBV, but although it was initially suggested that this was associated with reduced specific CTL in PBMC, later studies by YAO et al. (1985) failed to define a direct association. The increased incidence of EBV-associated lymphoproliferative disease with prolonged use of agents such as cyclosporin A and FK506, which suppress T cell activation, does, however, implicate mechanisms based on cellular immune responses. However, tumour regression often occurs within 1–2 weeks of stopping such medication, suggesting that CTL may not be directly involved in this process (NALESNIK 1991).

Another condition associated with outgrowth of EBV-transformed B cells is the X-linked lymphoproliferative syndrome (XLP) (PURTILLO 1991). This disease is characterised by a seemingly unique susceptibility of male members of affected families to EBV. This results in often fatal infectious mononucleosis, hypogammaglobulinaemia or malignant lymphoma, with 60% of males dying by 10 years of age. A defect has been identified located at Xq25–26 (WYANDT et al. 1989). A number of EBV-specific defects have been identified in these individuals, including a failure of NK and $CD8^+$ CTL responses to EBV-transformed B cells. However, the nature of the disease phenotype is still unclear, as a number of other defects are reported, including an early susceptibility to a variety of other viruses (for example, 5% had measles pneumonitis), as well as a failure to mount an efficient IgM response with immunoglobulin class switching to experimental challenge with ϕX174 (PURTILLO et al. 1989). Thus, although the failure to generate an effective $CD8^+$ CTL response may be associated with the development of EBV-associated lymphoproliferation, this does not occur in isolation, even in this genetically defined disorder.

Defining the precise role of $CD8^+$ CTL in γ-herpesvirus infections has been hampered by the absence of an animal model. The cottontop tamarind colony established in Bristol, UK, is a unique resource: the animals develop widespread lymphoma on exposure to EBV (SHOPE et al. 1973), but they are a protected species, precluding their widespread use as an experimental model. Recently, a murine herpesvirus with many of the features of EBV infection has been described by Nash and colleagues (SUNIL-CHANDRA et al. 1992a, b) and this may provide a model system where pathogenesis can be investigated in more detail.

6.2 Specificity for Viral Gene Products

Initial characterisation of the virus products recognised by these CTL was hampered by the failure to serologically define which virus product was involved. The concentration on membrane-associated proteins led to the circular definition of the target structure as LYDMA (lymphocyte-determined membrane antigen) that was present on autologous EBV-transformed cell lines (RICKINSON et al. 1981). However, as firstly influenza-specific CTL and then other virus-specific CTL were shown to recognise nuclear antigens, and then peptide fragments of these antigens associated with MHC class I, attention focused on the range of viral proteins expressed in EBV-transformed B cells in vitro. Less than 5% of EBV-transformed B cells express viral proteins associated with the lytic cycle of replication, and yet BCL are killed by EBV-specific $CD8^+$ CTL. Thus, a detailed examination of latency-associated proteins as targets for CTL has been pursued.

EBV-transformed B cells are known to express six viral nuclear proteins (EBNA), three membrane proteins (latent membrane proteins, LMP) and two small EBV-encoded RNAs (EBER; SAMPLE and KIEFF 1991). These proteins subserve a number of functions involved with maintenance of episomal replication and transmission of the genome (EBNA-1), immortalisation of B cells and alteration of host cell gene expression (EBNA-2), and inhibition of terminal differentiation of epithelial cells (LMP-1).

The mechanisms involved in immortalisation of B cells have not been fully elucidated. After binding of virus to the C3d (CD21) receptor, there is increased expression of CD23 (a B cell-derived B cell growth factor) with partial activation of the infected B cells (GORDON et al. 1986). EBNA-2 and EBNA-leader protein (EBNA-LP) are expressed after 8–12 h, with the other EBNAs and LMP after 32 h (ALFIERI et al. 1991). These latter changes are coincident with increased CD21 and CD23, together with increased c-*myc* expression. The EBER are not maximally expressed until 70 h post-infection. EBNA-2 is crucial for transformation; EBNA-2-deleted virus does not immortalise B cells, but this capacity is restored by replacing the gene

(HAMMERSCHMIDT and SUGDEN 1989). EBNA–LP, whilst not essential for immortalisation, augments the capacity of EBNA-2^+ virus to mediate this function (ROONEY et al. 1989). These observations based on studies with a mutant virus are consistent with the view that the other latency-associated viral proteins may also have an important role, and the function of EBNA-2 may be to activate expression of these and host cell genes to favour continued growth. This may be achieved by EBNA-2, which with LMP-1 protects infected cells from natural apoptosis (GREGORY et al. 1991).

Whilst attention focused on possible membrane-associated antigens that may induce a CTL response, the first suggestion was that LMP-1 may act as a CTL target (THORLEY-LAWSON and ISRAELSOHN 1987). In this study, a sequence of the LMP protein was able to activate a CTL response, but was not recognised as a peptide fragment in the conventional manner. Killing of double-transfected P815 murine tumour cells expressing HLA A11 and LMP was observed (MURRAY et al. 1988); however, the same group subsequently found that the determinant recognised was associated with the murine H-2 K^d, and not the transfected HLA gene (MURRAY et al. 1990). Recently, MOSS et al. (1991) have reported an epitope for CTL in LMP using vaccinia-infected target cells.

The definition of target EBV latent gene proteins for specific CTL has been aided by the identification of two distinct types of EBV: A and B. They are separated on the basis of their differences in the *Bam* H1 WYH and *Hind*III E regions—corresponding to sequence divergence in EBNA 2, 3, 4 and 6. The EBV type B proteins are both antigenically distinct and are less efficient in B cell immortalisation (RICKINSON et al. 1987). When EBV regression assays were performed using type A- or B-EBV, most donors had equivalent responses, but two subjects were identified in whom the response to type A virus was stronger than to type B, suggesting that epitopes specific to type A predominated in these individuals (BURROWS et al. 1990). A series of experiments using type A-specific CTL and type B-transformed autologous target cells, infected with vaccinia recombinants, demonstrated EBNA-2 (epitope located in the N-terminal 100 amino acids) and EBNA-3a (MURRAY et al. 1990) specificity. Peptides within EBNA-2 (DTPLIPLTIF—HLA A2; APOLLANI et al. 1992), EBNA-3 (FLRGRAYGI—HLA B8; BURROWS et al. 1990, 1992; MISKO et al. 1991) and EBNA-6 (EENLLDFVRF—HLA B44; APOLLANI et al. 1992; KEHVIQNAFRK—HLA A11), defined on the basis of sequence divergence between type A and B virus, but also by conformation with predictive algorithms (DELISI and BERZOFSKY 1985; ROTHBARD and TAYLOR 1988), are recognised by appropriately restricted specific CTL lines and clones (MOSS et al. 1991). Anti-IgM-activated B cell blasts infected with vaccinia recombinants expressing the EBV latent genes were also used as target cells, with CTL clones generated from autologous B cell lines (B95.8 transformed) to define B-type epitopes (MOSS et al. 1991). It is of particular interest that to date no CTL epitopes have been identified in EBNA-1, especially as Burkitt's lymphoma cells, which express only this EBNA protein

(ROWE et al. 1986), are relatively resistant to CTL lysis. Whether this is due to lack of appropriate virus antigen and/or reduced MHC class I, intercellular adhesion molecule (ICAM) and LFA-3 expression (GREGORY et al. 1988) is unknown (see below).

An important question in relation to the possible role of CTL in surveillance against persistent virus infection is whether selection of variant virus in vivo may occur by mutation in the CTL epitope. Such escape mutants are well described with antibody responses in lentivirus infection, and such a CTL escape mechanism has been proposed in HIV infection (PHILLIPS et al. 1991). APOLLONI et al. (1992) examined the sequence variation in the type A EBNA-3 (FLRGRAYGI) and EBNA-6 (EENLLDFVRF) epitopes in a number of clinical isolates (nasopharyngeal carcinoma, infectious mononucleosis, healthy donors) and laboratory strains. The EBNA-6 epitope was conserved in all viruses studied, but there was considerable variation in the EBNA-3 epitope. This epitope was not found in type B virus, but is conserved between even geographically distant type A viruses. B95.8 virus (maintained in vitro in marmoset cells) has an intermediate type (C-terminal isoleucine—leucine). This single change resulted in a 15-fold reduction in efficiency of the peptide as a CTL target. Systematic replacement of each amino acid in the FLRGRAYGL peptide resulted in only five peptides having a greater capacity to sensitize target cells for CTL lysis. From these observations it has been calculated that the CTL clone LC13 has a specificity for its target peptide of 1 in 4.7×10^{10}—a specificity greater than that for antibody–antigen interaction (BURROWS et al. 1992).

The question remains as to whether CTL recognise the structural and lytic pathway gene products of EBV. By analogy with CTL responses to other herpesviruses, it is probable that such a population of CTL precursors (CTLp) is present in PBMC. Various attempts have been made with individual proteins; POTHEN et al. (1991) induced T cell proliferation in vitro with the p17 and p50 components of the early antigen complex, but only $CD4^+$ T cell lines were established. As with all such studies, the difficulty may well lie with the inability to present the appropriate antigen in an autologous stimulator cell to induce specific $CD8^+$ CTL in vitro.

$CD4^+$ CTL restricted through MHC class II have been identified against EBV (MOSS et al. 1988; MISKO et al. 1984), including some with specificity for a number of structural proteins such as gp340. However, the role of such cells in vivo in the case of EBV, as with the other herpesviruses, is unknown. In seven seropositive subjects, the CTL frequency and MHC restriction was estimated, but no direct CD4 frequency could be estimated (BOURGAULT et al. 1991). Furthermore, during primary EBV mononucleosis, $CD8^+$ CTL were present but no $CD4^+$ CTL were detected in antibody-blocking experiments (ENSSLE and FLEISCHER 1990). However, the observations that long-term lines and clones may be detected raises the question as to whether the in vitro conditions used to generate class II-restricted responses in these studies might prejudice the outgrowth of such CTL. In both these studies EBV-transformed B cell lines were used to stimulate the cultures; thus, only a minority of the stimulator cells would have expressed structural proteins that may predominantly active CD4$^+$ responses. The assay protocol may therefore favour the detection of latency associated protein-specific CTL.

It is possible that CD4$^+$ CTL may be important in the very early response to EBV. When naive cells from EBV-seronegative donors were co-cultured with EBV-transformed B cell lines, surprisingly CD4$^+$ CTL, including type A- and B-EBV-specific CTL, were generated (MISKO et al. 1991), whereas in most such attempts with other herpesviruses no response is observed (TORPEY and LINDSLEY 1989; BORYSIEWICZ et al. 1983; HICKLING et al. 1987; YASUKAWA et al. 1989).

The nature of the restricting element in some EBV responses has also been difficult to determine. CHEN et al. (1989), using HLA-loss variants of the EBV-transformed cell line 721, which has lost HLA A and B determinants, yet retains HLA C molecules, have shown that this class I determinant may also serve as a restriction element. SVEDMYR and JONDAL (1975) described a population of EBV-specific, yet not MHC-restricted, CTL in PBMC of patients with EBV mononucleosis, which have not been detected in other herpesvirus infections. STRANG and RICKINSON (1987) re-examined this response and suggested that in the polyclonal activation of T and B cells that ensues following primary EBV infection, CTL with multiple MHC specificities are generated, which could explain the original 1975 findings.

7 CTL in Human Cytomegalovirus Infection

7.1 Asymptomatic Persistent Infection

As for detection of CTL in other human persistent virus infections, BORYSIEWICZ et al. (1983), used a secondary in vitro stimulation protocol which co-cultured PBMC with autologous HCMV-infected fibroblasts in the presence of exogenous IL-2. Using this approach, HCMV-specific CTL were detected in bulk cultures of PBMC, the requirement of cell-associated virus rather than free virus for generation of CD8+ CTL was confirmed, and a predilection of effector CTL for non-structural HCMV proteins identified (BORYSIEWICZ et al. 1983, 1988). Subsequently, this approach has been used by others to confirm these observations, and it was reported that recent clinical isolates of HCMV may suppress such CTL (SCHRIER and OLDSTONE 1986). A modification of this in vitro protocol, which enables the detection of HCMV-specific CTL after 6 days culture, has been described (LAUBSCHER et al. 1988).

Bulk culture of CTL has major limitations. Although it allows the detection of a response, this is in essence a qualitative measure, as in vitro amplification does not allow the magnitude of ^{51}Cr release to be directly related to the number of CTLp initiating a culture. Thus, BORYSIEWICZ et al. (1988) used limiting dilution culture with split-well analysis to estimate the frequency and stage specificity of HCMV CTL. In PBMC of normal HCMV-seropositive subjects, such CTL are present at 1/5000 to 1/50000 PBMC. This frequency is comparable to that described for EBV CTL (see above), but much greater than that for VZV (1/60000), even in subjects with recent clinical reactivation (HICKLING et al. 1987).

These studies initially confirmed the preferential lysis of HCMV-infected cells treated with phosphonoformate to inhibit de novo production of HCMV structural proteins (BORYSIEWICZ et al. 1988). However, the limiting dilution technique was also adapted to estimate the relative number of CTL against individual HCMV gene products (BORYSIEWICZ et al. 1988), as discussed below.

HCMV-specific $CD4^+$, MHC class II-restricted CTL have also been described (LINDSLEY et al. 1986). Such CTL were shown to kill autologous HCMV-infected monocytes, suggesting that they may have a role in controlling the infection in monocytes, which can be infected by the virus and which may indeed be an important site of virus persistence in vivo (TAYLOR-WIEDEMAN et al. 1991). However, it is difficult to pursue these studies in the detail that has proved possible for MHC class I-restricted CTL responses, as the only readily available target cell that supports a full lytic cycle of HCMV replication is the human diploid fibroblast, which does not constitutively express MHC class II determinants, even when HCMV infected.

7.2 Disease and Immunosuppression

Initial studies of the CTL response against HCMV focused on direct detection of these effector cells in PBMC without requirement for in vitro stimulation. QUINNAN et al. (1982) detected MHC class I-restricted responses in PBMC of transplant recipients in a direct CTL assay using HCMV-infected frozen target cells of known MHC type. They found that the absence of this response, in the presence of clinical HCMV infection, was associated with a poor prognosis and recommended the assay for direct clinical use. However, BRENING et al. (1984) found that the assay was difficult to interpret in clinical use, and no CTL could be detected by this approach in asymptomatic persistently HCMV-infected subjects (QUINNAN et al. 1982, 1984). However, this approach was used to detect the presence of HCMV CTL in renal transplant recipients and in patients with acquired immunodeficiency syndrome (AIDS) (ROOK et al. 1985).

HCMV may infect a number of different cell types in vivo, in contrast to the restriction of in vitro replication to human diploid fibroblasts. In normal subjects such cells probably include epithelial cells, monocytes/macrophages, and bone marrow precursor cells (MINTON et al. 1993); in HCMV disease, other cells including polymorphonuclear leukocytes may also be infected. In addition, the state of cell differentiation and division may itself determine whether virus replication can proceed; monocytes cultured in vivo with T cell supernatant medium or in the presence of corticosteroids permit full virus replication in vitro with detection of released infectious particles (LATHEY and SPECTOR 1991). Thus, the potential range of infected target cells in vivo may well be greater than that so far studied in vitro.

To date, studies with MCMV in Balb/c mice have suggested that CTL are important in protection against primary infection and in recovery from established infection after adoptive transfer (REDDEHASE et al. 1987). In man, such studies are necessarily correlative. QUINNAN'S original observations suggest that these cells may be directly protective in bone marrow transplant recipients. We have performed quantitative studies to determine the CTLp frequency in renal and heart lung transplant recipients. With administration of immunosuppression, CTLp were low, but in survivors of HCMV pneumonitis the CTLp frequency increased (ALP et al., manuscript submitted). In the convalescent phase of HCMV pneumonitis, we have to date been unable to recover HCMV-specific CTL in bronchoalveolar lavage lymphocytes (R. Smyth, unpublished observations). Furthermore, where cytotoxicity of HCMV-infected cells by bronchoalveolar lymphocytes from bone marrow transplant recipients with HCMV pneumonitis has been observed, this cytotoxicity was abrogated by anti-Leu-11b, indicating that NK cells rather than specific CTL were involved (BOWDEN et al. 1988).

Such observations bear on the question of the possible role of CTL in the pathogenesis of organ-specific HCMV-induced injury. On the basis of studies in MCMV and the low frequency of HCMV pneumonitis (as compared with HCMV retinitis) in HIV-infected patients, who may have difficulty in mounting an effective $CD8^+$ CTL response, Grundy et al. (1987) suggested that HCMV-induced lung injury was immunopathogenically mediated. However, in a recent study CARMICHAEL et al. (1993) have found that although the HIV-1-specific CTL response in PBMC is reduced with the development of stage IV disease and a low CD4 count, the CTL response to EBV is preserved with normal numbers of CTLp until very late in the disease. Were the same to apply for HCMV-specific CTL, then lack of $CD8^+$ CTL might not be the explanation for the difference in HCMV disease observed in AIDS patients.

The temporal relationship between the development of HCMV-specific CTL responses following bone marrow transplantation and HCMV-related complications has been re-examined by REUSSER et al. (1991). A total of 20 bone marrow transplant recipients were studied for the presence of HCMV-specific CTL in bulk culture after in vitro stimulation with HCMV-infected fibroblasts. Of the 20 donors providing the transplant to the recipients, 17 generated such responses; of the 20 recipients, ten developed a HCMV-specific CTL response within 2 months of transplantation and only one developed significant HCMV disease. Six of the ten who failed to generate such CTL developed fatal HCMV pneumonitis. Again, this study suggests that the $CD8^+$ CTL response is protective against pneumonitis and not harmful to the host. Furthermore, it provides a rationale for considering adoptive transfer therapy against this disease in the transplant recipient; such studies are in progress in Seattle (RIDDELL et al. 1992).

7.3 Specificity for Viral Gene Products

Using vaccinia recombinants encoding the HCMV 72-kDa IE-1 protein and a structural protein, glycoprotein B, it was found that a high proportion of CTL clones recognised the IE product (20%–60%) as compared with glycoprotein B (<4%; BORYSIEWICZ et al. 1988). The fine specificity of the 72-kDa IE-1 protein-specific CTL restricted by HLA B13 has been determined and the peptide located in exon 4 of IE-1 (ALP et al., manuscript submitted). Interestingly, this region has no overlap peptides inducing CD4 T cell proliferation (ALP et al. 1991). CTL epitopes have also been identified in glycoprotein B (613-PSLKIFIAGNSAYEY-V-628) in association with HLA A2 (PARKER et al. 1992). Recently, using a range of vaccinia recombinants expressing truncates of glycoprotein B, we have found multiple CTL epitopes dependent on the MHC type of the subject tested, thereby suggesting that no immunodominant CTL region in this protein can be identified (unpublished observation).

RIDDELL et al. (1991) have identified an MHC class I-restricted CTL response against a HCMV structural protein, although the precise specificity remains to be identified. They employed polyclonal activation of PBMC (RIDDELL and GREENBERG 1990) with limiting dilution cloning to isolate HCMV-specific CTL. They have also identified CTL specific for HCMV 72-kDa IE-1 response in bulk culture and individual clones (RIDDELL et al. 1991). However, they also found that HCMV-infected actinomycin D-treated target cells were killed, implying that there was no requirement for de novo viral protein synthesis to sensitize the cell (as would be the case with IE-1-specific CTL) and suggesting that a structural protein entering with the input virus could sensitize the infected cell to MHC class I-restricted CTL lysis (RIDDELL et al. 1991).

8 Conclusions on the Role of Herpes Virus-Specific CTL In Vivo

8.1 Relationship Between CTL Activity and Virus Reactivation

No definite conclusion can be reached on the role that CTL play in reactivation of latent infection. It seems intrinsically unlikely that they play any role in inhibiting the actual induction of IE gene expression from hitherto latent virus and more likely that they may control reactivation episodes once they have occurred, perhaps preventing reactivation from achieving clinical significance by containing it at local sites. Indeed, it can be argued that the maintenance of the relatively high CTL frequencies observed in asymptomatic herpes-virus carriers may be secondary to continual or frequent restimulation in vivo, consequent on expression of viral genes due to reactivation.

There is a need for more studies of CTL using the cell types naturally carrying virus in vivo as targets.

8.2 Evidence for Protective Role of CTL

It is similarly difficult to draw firm conclusions on the protective role of CTL in human herpesvirus infections. Their protective role is inferred from the increased morbidity from herpesvirus infections in subjects with impairment of T cell responses, although other effector T cell functions are almost invariably concomitantly impaired. Any more definite conclusions are only likely to result from the selective induction of CTL by immunisation or possibly by the human adoptive transfer experiments referred to below in the context of HCMV.

8.3 Possible Role in Immunopathology

There are experimental models in which CTL appear to play a pathogenic role, the best known being the lymphocytic choriomeningitis virus model, in which CTL produce cerebral damage following intracerebral inoculation of virus in immunocompetent adult mice (BUCHMEIER et al. 1980). Is there any immunopathogenic role for CTL in herpesvirus infections? As noted above, it has been suggested that HCMV-specific CTL may be responsible for the pathogenesis of HCMV pneumonitis (GRUNDY et al. 1987). However, this is speculative, based on the relative rarity of this condition in AIDS patients and its peak incidence in bone marrow transplant recipients at the time of reconstitution. There is as yet no direct evidence in man to support this interesting suggestion.

8.4 Role Relative to Other Immune Responses

The concentration on CTL in this review does not imply that other immunological mechanisms are of little importance in resistance to herpesvirus infections. Antibody is likely to play a major role in resistance to primary infection. There is evidence that the administration of CMV-immune globulin to seronegative transplant recipients does reduce the incidence of severe HCMV disease (SNYDMAN 1991). However, given that reactivation and disseminated infection may occur in the face of circulating antibody, it seems less likely that it is of primary importance in controlling reactivation.

There is increasing evidence for some redundancy in the immune system. In experimental models, the depletion of $CD8^+$ cells can lead to $CD4^+$ T cells controlling MCMV infection (JONJIC et al. 1989). Congenital absence of $CD8^+$ T cells, as in "knockout" mice for β_2-microglobulin which have no class I MHC, appears to increase the likelihood of compensatory antiviral $CD4^+$ CTL responses developing in the case of LCMV (MULLER et al. 1992).

9 The Induction of CTL In Vivo by Immunisation

There are currently candidate vaccines in development for each of the human herpesviruses (except HHV-6). Two of these, which are live attenuated vaccines, have been administered to human subjects quite extensively (for VZV and CMV); there are also a number of subunit vaccines in development for each of the viruses, and some of these have been used in experimental models. All of these candidate vaccines have been developed with the primary purpose of inducing neutralising antibody to envelope proteins; however, given the recognition of the importance of CTL in controlling persistent infections, there is interest in whether these vaccines can induce virus-specific CTL. There have also been attempts to adoptively transfer human herpesvirus specific CTL.

9.1 Induction of CTL by Vaccines

HSV vaccines are in development, based on recombinant glycoproteins such as gB, and two such vaccines are now entering trials for the prevention of reactivation episodes in people with recurrent genital herpes. However, as yet there have not been any reported studies of the immune responses they induce in humans (ALLEN and HITCHCOCK 1991).

VZV vaccines based on recombinant glycoproteins are also in development, but a live attenuated candidate vaccine—the Oka strain vaccine—has been quite extensively used for some time. The vaccine has been shown to provide a degree of immunity from primary VZV infection in previously seronegative subjects who are immunosuppressed and in normal individuals at particular risk of VZV infection (GERSHON and STEINBERG 1989, 1990). This vaccine is reported to induce VZV-specific CTL of both $CD4^+$ and $CD8^+$ phenotype; such CTL show specificity for the IE-62 protein and for gpI, with frequencies comparable to those induced by natural infection (ARVIN 1992).

The vaccine enhances proliferative T cell frequencies in seropositive elderly subjects (from 1/68000 before vaccination to 1/40000) as effectively as it does in younger subjects (LEVIN et al. 1992).

EBV, a candidate vaccine based on gp340, has been used in cottontop tamarins to prevent EBV-induced lymphomas (FINERTY et al. 1992; MORGAN 1991). Its ability to induce CTL in this experimental model is not known and it has not so far been given to humans, although at the time of writing it seems that clinical trials may start soon. A particular target population for an EBV vaccine would probably be those in the Far East and China at risk of nasopharyngeal carcinoma.

CMV vaccines have attracted interest because of the severity of the clinical disease which HCMV produces in transplant populations and in the infected neonate. Potential target populations for a vaccine are seronegative pregnant mothers and transplant recipients. The only CMV vaccine which has been used to any extent in humans is a live Towne strain virus, which is presumed to be attenuated, although there is no marker for attenuation. This vaccine is reported to give a level of protection to seronegative transplant recipients comparable to that given by previous infection with CMV, the risk of CMV disease in vaccinated subjects approximating to that in unvaccinated CMV seropositive subjects (reviewed in GONCZOL and PLOTKIN 1990). The immune response this vaccine induces in recipients includes induction of proliferative T cell responses, but there is no description of the induction of CTL by the vaccine. Subunit CMV vaccines based on CMV glycoproteins are now in development, although none have yet been subjected to human trials.

9.2 Adoptive Transfer of Human CTL

Although it might be thought impracticable to apply on any scale, adoptive transfer of human virus-specific CTL is being seriously explored by at least one group. HCMV-specific CTL $CD8^+$ clones were grown in vitro from seropositive bone marrow donors by the system described above and then administered to the recipients of bone marrow transplants from the same donors during the period when they are at maximum risk of CMV infection. The Seattle group have reported that after transfer of increasing numbers of these CTL (from 10^8 to 10^9 cells in four doses over 4 weeks), CMV-specific cytotoxicity could be detected in the PBMC of recipients for up to 1 month (RIDDELL et al. 1992). It will be of great interest to see whether this adoptive transfer confers protection from clinically significant CMV disease in the recipients. However, such protection may be difficult to demonstrate, as a large number of subjects might well be needed to show a benefit if all receive the current CMV prophylaxis regimes which have already been shown to diminish the incidence of CMV disease very significantly (GOODRICH et al. 1991). It would seem unlikely that this approach will be pursued for herpes-virus infections other than CMV.

9.3 Other Approaches

Other approaches to the induction or boosting of CTL immunity might be feasible, but are largely speculative. There are no reports of the effect of LAK cells, as used in the therapy of cancers, for treating virus infections in immunosuppressed individuals, although these could in some respects be regarded as a necessary control for the effects of the adoptive immunotherapy approach described above. The use of "post-infective" immunisation by administering vaccines to seropositive subjects at risk, in order to boost their immunity including CTL, might also be worth considering. A trial of this approach has already been reported in HIV infection, although there is no convincing evidence of clinical benefit as yet (REDFIELD et al. 1991).

10 Conclusions

The evidence for the role of CTL in animal models of virus infections, coupled with the availability of an in vitro functional assay, has led to numerous studies of their possible role in human herpesvirus infection. CTL can be demonstrated in all human herpesvirus infections, but so far evidence for their protective role is, perhaps inevitably, indirect at best. Given the current level of interest in both the mechanisms of induction of CTL and the immunological control of persistent virus infections (additionally driven by work on HIV infection), we should learn considerably more about the role of CTL in herpesvirus infections over the next few years.

References

Alfieri C et al. (1991) Early events in Epstein-Barr virus infection of human B lymphocytes. Virology 181: 595

Allen WP, Hitchcock PJ (eds) (1991) Herpes simplex virus vaccine workshop 1989. Rev Infect Dis 13 [Suppl 11]: S891–S979

Alp N et al. (1990) Automation of limiting dilution cytotoxicity assays. J Immunol Methods 129: 269–276

Alp N et al. (1991) The fine specificity of cellular immune responses in man to human cytomegalovirus immediate early 1 protein. J Virol 65: 4812–4820

Apollani A et al. (1992) Sequence variation of cytotoxic T cell epitopes in different isolates of Epstein-Barr virus. Eur J Immunol 22: 183–189

Arvin AM (1992) Cell-mediated immunity to varicella-zoster virus. J Infect Dis 166 [Suppl 1]: S35–S41

Arvin AM et al. (1991) Equivalent recognition of a varicella-zoster virus immediate early protein (IE62) and glycoprotein 1 by cytotoxic T lymphocytes of either CD^4 or $CD8^+$ phenotype. J Immunol 146: 257–264

Baer R et al. (1984) DNA sequence and expression of the B95-8 Epstein-Barr virus genome. Nature 310: 207–211

Beck S, Barrell BG (1988) HCMV encodes a glycoprotein homologous to MHC class-1 antigen. Nature 331: 269–272
Bergen RE et al. (1991) Human T cells recognize multiple epitopes of an immediate early/tegument protein (IE62) and glycoprotein I of varicella-zoster virus. Viral Immunol 4: 151–166
Bonneau RH, Jennings SR (1988) Modulation of acute and latent herpes simplex virus infection in C57BL/6 mice by adoptive transfer of immune lymphocytes with cytolytic activity. J Virol 63: 1480–1484
Borysiewicz LK et al. (1983) Requirements for in vitro generation of human cytomegalovirus specific cytotoxic T cell lines. Eur J Immunol 13: 804–809
Borysiewicz LK et al. (1988a) Human cytomegalovirus-specific cytotoxic-T-cells: their precursor frequency and stage specificity. Eur J Immunol 18: 269–275
Borysiewicz LK et al. (1988b) Human cytomegalovirus specific cytotoxic T cells recognise the 72kD immediate early protein and glycoprotein B in recombinant vaccinia viruses. J Exp Med 168: 919–931
Bourgault I et al. (1991) Limiting dilution analysis of the HLA restriction of EBV-specific cytolytic T lymphocytes. Clin Exp Immunol 84: 501–507
Bowden RA et al. (1985) Lysis of varicella zoster virus infected cells by lymphocytes from normal humans and immunosuppressed pediatric leukaemia patients. Clin Exp Immunol 60: 387–395
Bowden RA et al. (1988) Increased cytotoxicity against cytomegalovirus-infected target cells by bronchoalveolar lavage cells from bone marrow transplant recipients with cytomegalovirus pneumonitis. J Immunol Methods 158: 773–779
Brening MK et al. (1984) Human and murine MHC-restricted cytotoxic lymphocyte responses to CMV infection. Birth Defects 20: 375–379
Buchmeier MJ et al. (1980) The virology and immunobiology of lymphocytic choriomeningitis virus infection. Adv Immunol 30: 275–331
Bukowski JF et al. (1985) Adoptive transfer studies demonstrating the antiviral effect of natural killer cells in vivo. J Exp Med 161: 40–52
Burrows SR et al. (1990a) An Epstein-Barr virus-specific cytotoxic T cell epitope in EBV nuclear antigen 3 (EBNA3). J Exp Med 171: 345–349
Burrows SR et al. (1990b) Patterns of reactivity of Epstein-Barr virus specific T cells in type A donor cultures after reactivation with A- or B-transformants. Cell Immunol 127: 47
Burrows SR et al. (1992) The specificity of recognition of a cytotoxic T lymphocyte epitope. Eur J Immunol 22: 191–195
Carmichael A et al. (1993) Quantitative analysis of the human immunodeficiency virus type 1 (HIV-1)-specific cytotoxic T lymphocyte (CTL) response at different stages of HIV-1 infection: differential CTL responses to HIV-1 and Epstein-Barr virus in late disease. J Exp Med 177: 249–256
Chee MS et al. (1990) Analysis of the protein-coding content of the sequence of human cytomegalovirus strain AD169. In: McDougall JK (ed) Cytomegaloviruses. Springer, Berlin Heidelberg New York, pp 125–170 (Current topics in microbiology and immunology, vol 154)
Chen BP et al. (1989) Restriction of Epstein-Barr virus specific cytotoxic T cells by HLA-A, -B and -C molecules. Hum Immunol 26(2): 137–147
Cunningham AL, Noble JR (1989) Role of keratinocytes in recurrent herpetic lesions. Ability to present herpes simplex virus antigen and act as targets for T lymphocytotoxicity in vitro. J Clin Invest 83: 490–496
Cunningham AL et al. (1985) Evolution of recurrent herpes simplex lesions: an immunohistologic study. J Clin Invest 75: 226–233
Davison AJ, Scott JE (1986) The complete DNA sequence of varicella-zoster virus. J Gen Virol 67: 1759–1816
DeLisi C, Berzofsky JA (1985) T cell antigen sites tend to be amphipathic structures. Proc Natl Acad Sci USA 81: 7632
Diaz PS et al. (1989) T lymphocyte cytotoxicity with natural varicella-zoster virus infection and after immunization with live attenuated varicella vaccine. J Immunol 142: 636–641
Enssle KH, Fleischer B (1990) Absence of EBV specific, HLA class II-restricted $CD4^+$ cytotoxic T cells in infectious mononucleosis. Clin Exp Immunol 79: 409–415
Finerty S et al. (1992) Protective immunization against Epstein-Barr virus-induced disease in cottontop tamarins using the virus envelope glycoprotein gp340 produced from a bovine papillomavirus expression vector. J Gen Virol 73: 449–453

Garcia-Blanco MA, Cullen BR (1991) Molecular basis of latency in pathogenic human viruses. Science 254: 815–820
Gershon AA, Steinberg SP (1989) Varicella vaccine collaborative study group of the National Institute of Allergy and Infectious Diseases. Persistence of immunity to varicella in children with leukemia immunized with live attenuated varicella vaccine. N Engl J Med 320: 892–897
Gershon AA, Steinberg S (1990) Live attenuated varicella vaccine: protection in healthy adults in comparison to leukemic children. J Infect Dis 161: 661–666
Gonczol E, Plotkin S (1990) Progress in vaccine development for prevention of human CMV infection. In: McDougall JK (ed) Cytomegaloviruses. Springer, Berlin Heidelberg New York, p 255 (Current topics in microbiology and immunology, vol 154)
Goodrich JM et al. (1991) Early treatment with ganciclovir to prevent CMV disease after bone marrow transplantation. N Engl J Med 325: 1601–1607
Gordon J et al. (1986) Control of human B-lymphocyte replication: transforming Epstein-Barr virus exploits three distinct viral signals to undermine three separate control points in B-cell growth. Immunology 58: 591
Gregory CD et al. (1988) Down-regulation of cell adhesion molecules LFA-3 and ICAM-1 in Epstein-Barr virus positive Burkitt's lymphoma underlies tumour cell escape from virus-specific T cell surveillance. J Exp Med 167: 1811–1824
Gregory CD et al. (1991) Activation of Epstein-Barr virus latent genes protects human B cells from death by apoptosis. Nature 349: 612–614
Grundy JE et al. (1987) Is cytomegalovirus interstitial pneumonitis in transplant recipients an immunopathological condition? Lancet 2: 996–999
Hagemeier C et al. (1992) The 72K IE1 and 80K IE2 proteins of human cytomegalovirus independently transactivate the c-fos, c-myc and hsp70 promoters via basal promoter elements. J Virol 73: 2385–2393
Hammerschmidt W, Sugden B (1989) Genetic analysis of the immortalising functions of Epstein-Barr virus in human B lymphocytes. Nature 340: 393–395
Hardy I et al. and the Varicella Vaccine Collaborative Study Group (1991) The incidence of Zoster after immunization with live attenuated varicella vaccine—a study in children with leukemia. N Engl J Med 325: 1545–1550
Hayward AR, Herberger M (1987) Lymphocyte responses to varicella zoster virus in the elderly. J Clin Immunol 7: 174–178
Hayward AR et al. (1986a) Cellular interactions in the lysis of varicella zoster virus infected human fibroblasts. Clin Exp Immunol 63: 141–146
Hayward AR et al. (1986b) Specific lysis of varicella zoster virus-infected B lymphoblasts by human T cells. J Virol 58: 179–184
Hayward A et al. (1991) Varicella-Zoster virus-specific immunity after herpes zoster. J Infect Dis 163: 873–875
Hickling J et al. (1987) Varicella zoster virus specific cytotoxic T lymphocytes (Tc): detection and frequency analysis of HLA class I restricted Tc in human peripheral blood. J Virol 61: 3463–3469
Ibanez CE et al. (1991) Human cytomegalovirus productively infects primary differentiated macrophages. J Virol 65: 6581–6588
Jonjic S et al. (1988) A non-structural viral protein expressed by a recombinant vaccinia virus protects against lethal cytomegalovirus infection. J Virol 62: 1653–1658
Jonjic S et al. (1989) Site-restricted persistent cytomegalovirus infection after selective long-term depletion of CD4+ T lymphocytes. J Exp Med 169: 1199–1212
Koszinowski UH et al. (1990) Cellular and molecular basis of the protective immune response to cytomegalovirus infection. In: McDougall JK (ed) Cytomegaloviruses. Springer, Berlin Heidelberg New York, pp 189–220 (Current topics in microbiology and immunology, vol-154)
Lathey JL, Spector SA (1991) Unrestricted replication of human cytomegalovirus in hydrocortisone-treated macrophages. J Virol 65: 6371–6375
Laubscher A et al. (1988) Generation of human cytomegalovirus-specific cytotoxic T lymphocytes in a short term culture. J Immunol Methods 110: 69–77
Levin MJ et al. (1992) Immune response of elderly individuals to a live attenuated varicella vaccine. J Infect Dis 166: 253–259
Lindsley MD et al. (1986) HLA-DR-restricted cytotoxicity of cytomegalovirus-infected monocytes mediated by Leu-3-positive T cells. J Immunol 136: 3045–3051
Lucin P et al. (1992) Gamma interferon dependent clearance of CMV infection in the salivary gland. J Virol 66: 1977–1984

Martin S et al. (1988) Herpes simplex type-1 specific cytotoxic T cells recognise virus nonstructural proteins. J Virol 62: 2265–2273
Mester JC, Rouse BT (1991) The mouse model and understanding immunity to herpes simplex virus. Rev Infect Dis 13 [Suppl 11]: S935–945
Minton EJ et al. (1993) Human cytomegalovirus infection of the monocyte/macrophage lineage: bone marrow progenitors may act as a reservoir for the virus, but monocyte differentiation is required for viral gene expression (in press)
Misko IS et al. (1984) HLA DR-antigen associated restriction of EBV specific T-cell colonies. Int J Cancer 33: 239
Misko IS et al. (1991a) Cytotoxic T lymphocyte discrimination between type A Epstein-Barr virus transformants is mapped to an immunodominant epitope in EBNA3. J Gen Virol 72: 405–409
Misko IS et al. (1991b) Composite response of naive T cells to stimulation with the autologous lymphoblastoid cell line is mediated by CD4 cytotoxic T cell clones and includes an Epstein-Barr virus-specific component. Cell Immunol 132: 295–307
Morgan AJ (1991) Control of viral disease: the development of EBV vaccines. Springer Semin Immunopathol 13: 249–262
Moss DJ et al. (1978) Long-term T-cell mediated immunity to Epstein-Barr virus in man. I. Complete regression of virus-induced transformation in cultures of seropositive donor leukocytes. Int J Cancer 22: 662
Moss DJ et al. (1988) Cytotoxic T cell clones discriminate between A- and B-type Epstein-Barr virus transformants. Nature 331: 719–721
Moss DJ et al. (1991) Immune regulation of Epstein-Barr virus; EBV nuclear antigen as a target for EBV-specific T cell lysis. Springer Semin Immunopathol 13: 147–156
Muller D et al. (1992) LCMV-specific, class II-restricted cytotoxic T cells in B2-microglobulin-deficient mice. Science 255: 1576–1578
Murray RJ (1992) Identification of target antigens for the human cytotoxic T cell response to Epstein-Barr virus (EBV): implications for the immune control of EBV-positive malignancies. J Exp Med 176: 157–168
Murray RJ et al. (1988) Epstein-Barr virus-specific cytotoxic T cell recognition of transfectants expressing the virus-coded latent membrane protein LMP. J Virol 62: 3747–3755
Murray RJ et al. (1990a) Cross recognition of a mouse H2-peptide complex by human HLA-restricted cytotoxic T cells. Eur J Immunol 20: 659–664
Murray RJ et al. (1990b) Human cytotoxic T-cell responses against Epstein-Barr virus nuclear antigens demonstrated by using recombinant vaccinia viruses. Proc Natl Acad Sci USA 87: 2906–2910
Nalesnik MA (1991) Lymphoproliferative disease in organ transplant recipients. Springer Semin Immunopathol 13: 199–216
Nash AA et al. (1987) Different roles for the $L3T4^+$ and $Lyt2^+$ T cell subsets in the control of an acute herpes simplex virus infection of the skin and nervous system. J Gen Virol 68: 825–833
Parker KC et al. (1992) Sequence motifs important for peptide binding to the human MHC class molecule HLA A2. J Immunol 149: 3580–3587
Petersen CM et al. (1989) Immunosuppressive properties of electrophoretically "slow" and "fast" form alpha2-macroglobulin. Effects on cell-mediated cytotoxicity and (allo-) antigen-induced T cell proliferation. J Immunol 142: 629–635
Phillips RE et al. (1991) Human immunodeficiency virus genetic variation that can escape cytotoxic T cell recognition. Nature 354: 453–459
Pothen S et al. (1991) Human T-cell recognition of Epstein-Barr virus-associated replication antigen complexes. Int J Cancer 49: 656–660
Purtillo DT (1991) X-linked lymphoproliferative disease (XLP) as a model of Epstein-Barr virus-induced immunopathology. Springer Semin Immunopathol 13: 181–197
Purtillo DT et al. (1989) Detection of X-linked lymphoproliferative disease (XLP) using molecular and immunovirological markers. Am J Med 87: 421
Quinnan GV et al. (1982) Cytotoxic T cells in CMV infection. N Engl J Med 307: 7–13
Quinnan GV et al. (1984) HLA-restricted cytotoxic T lymphocytes are an early immune response and important defense mechanism in cytomegalovirus infections. Rev Infect Dis 6: 156–163
Reddehase MJ et al. (1984) The cytolytic T lymphocyte response to the murine cytomegalovirus. II. Detection of virus stage-specific antigens by separate populations of in vivo active cytolytic T lymphocyte precursors. Eur J Immunol 14: 56–61
Reddehase MJ et al. (1987) CD8-positive T lymphocytes specific for murine cytomegalovirus immediate early antigens mediate protective immunity. J Virol 61: 3102–3108

Redfield RR et al. (1991) A phase 1 evaluation of the safety and immunogenicity of vaccination with recombinant gp160 in patients with early HIV infection. N Eng J Med 324: 1677–1684
Reusser P et al. (1991) Cytotoxic T-lymphocyte response to cytomegalovirus after human allogeneic bone marrow transplantation: pattern of recovery and correlation with cytomegalovirus infection and disease. Blood 78: 1373–1380
Rickinson AB et al. (1981) Long-term cell-mediated immunity to Epstein-Barr virus. Cancer Res 41: 4216
Rickinson AB et al. (1987) Influence of the Epstein-Barr virus nuclear antigen EBNA2 on the growth phenotype of virus-transformed B cells. J Virol 61: 1310
Riddell SR, Greenberg PD (1990) The use of anti-CD3 and anti-CD28 monoclonal antibodies to clone and expand human antigen specific T cells. J Immunol Methods 128: 189–201
Riddell SR et al. (1991a) Class I MHC-restricted cytotoxic T lymphocyte recognition of cells infected with human cytomegalovirus does not require endogenous viral gene expression. J Immunol 146: 2795–2804
Riddell SR et al. (1991b) Cytotoxic T cells specific for cytomegalovirus: a potential therapy for immunocompromised patients. Rev Infect Dis 13 [Suppl 11]: S966–973
Riddell SR et al. (1992) Restoration of viral immunity in immunodeficient humans by the adoptive transfer of T cell clones. Science 257: 238–241
Roizman B (1990) Herpesviridae. In: Fields BN, Knipe DM (eds) Virology, Raven, New York, pp 1787–1794
Roizman B, Sears AE (1990) Herpes simplex viruses and their replication. In: Fields BN, Knipe DM (eds) Virology, Raven, New York, pp 1795–1841
Rook AH et al. (1985) Deficient HLA-restricted, cytomegalovirus specific cytotoxic T cells and natural killer cells in patients with the acquired deficiency syndrome. J Infect Dis 152: 627–630
Rooney C et al. (1989) Influences of Burkitt's lymphoma and primary B cells on latent gene expression by the non-imortalising P3J-HR-1 strain of Epstein-Barr virus. J Virol 63: 1531
Rothbard JB, Taylor WR (1988) A sequence pattern common to T cell epitopes. EMBO J 7: 93–100
Rowe DT et al. (1986) Restricted expression of EBV latent genes and T-lymphocyte detected membrane antigen in Burkitt's lymphoma cells. EMBO J 5: 2599
Rowe M et al. (1992) Three pathways of Epstein-Barr virus gene activation from EBNA1-positive latency in B lymphocytes. J Virol 66: 122–131
Sample C, Kieff E (1991) Molecular basis for Epstein-Barr virus induced pathogenesis and disease. Springer Semin Immunopathol 13: 133–146
Schmid DS (1988) The human MHC-restricted cellular response to herpes simplex virus is mediated by $CD4^+$, $CD8^-$ T cells and is restricted to the DR region of the MHC complex. J Immunol 140: 3610–3616
Schrier RD, Oldstone MBA (1986) Recent isolates of CMV suppress CMV-specific human leukocyte antigen-restricted cytotoxic T lymphocyte activity. J Virol 69: 127–132
Schwartz RH (1992) Costimulation of T lymphocytes: the role of CD28, CTLA-4, and B7/BB1 in interleukin-2 production and immunotherapy. Cell 71: 1065–1068
Sharp M et al. (1992) Kinetics and viral protein specificity of the cytotoxic T lymphocyte response in healthy adults immunized with live attenuated varicella vaccine. J Infect Dis 165: 852–858
Shope T et al. (1973) Malignant lymphoma in cotton topped marmosets after innoculation with Epstein-Barr virus. Proc Natl Acad Sci USA 70: 2487–2491
Snydman DR (1991) Prevention of cytomegalovirus-associated diseases with immunoglobulin. Transplant Proc XXIII: [3, Suppl 3]: 131–135
Stamminger T, Fleckenstein B (1990) Immediate-early transcription regulation of human cytomegalovirus. In: McDougall JK (ed) Cytomegaloviruses. Springer, Berlin Heidelberg New York, p 154 (Current topics in microbiology and immunology, vol 154)
Stanberry LR (1991) Evaluation of herpes simplex virus vaccines in animals: the guinea pig vaginal model. Rev Infect Dis 13 [Suppl 11]: S920–923
Strang G, Rickinson AB (1987) Multiple HLA class I-dependent cytotoxicities constitute the "non-HLA-restricted" response in infectious mononucleosis. Eur J Immunol 17: 1007–1013
Sunil-Chandra NP et al. (1992a) Virological and pathological features of mice infected with murine gammaherpesvirus 68. J Gen Virol 73: 2347–2356
Sunil-Chandra NP et al. (1992b) Murine gammaherpesvirus 68 establishes a latent infection in mouse B lymphocytes in vivo. J Gen Virol 73: 3275–3279

Svedmyr E, Jondal M (1975) Cytotoxic effector cells specific for B cell lines transformed by Epstein-Barr virus are present in patients with infectious mononucleosis. Proc Natl Acad Sci USA 72: 1622

Taylor-Wiedeman J et al. (1991) Monocytes as a major site of persistence of human cytomegalovirus in peripheral blood mononuclear cells. J Gen Virol 72: 2059–2064

Thomson BJ et al. (1991) The molecular and cellular biology of human herpesvirus-6. Rev Med Virol 1: 89–99

Thorley-Lawson DA, Israelsohn ES (1987) Generation of specific cytotoxic T cells with a fragment of the Epstein-Barr virus-encoded p63/latent membrane protein. Proc Natl Acad Sci USA 84: 5484

Tigges MA et al. (1992) Human $CD8^+$ herpes simplex virus-specific cytotoxic T-lymphocyte clones recognize diverse virion protein antigens. J Virol 66: 1622–1634

Torpey DJ, Lindsley MD (1989) HLA-restricted lysis of herpes simplex virus-infected monocytes and macrophages is mediated by $CD4^+$ and $CD8^+$ T lymphocytes. J Immunol 142: 1325–1332

Wilson A et al. (1992) Subclinical VZV viraemia, herpes zoster and T lymphocyte immunity to VZV antigens after bone marrow transplantation. J Infect Dis 165: 119–126

Wyandt HE et al. (1989) Chromosomal deletion of Xq25 in an individual with X-linked lymphoproliferative disease. Am J Hum Genet 33: 426

Yao QY et al. (1985) In vitro analysis of the Epstein-Barr virus: host balance in long-term renal allograft recipients. Int J Cancer 35: 43

Yasukawa M, Zarling JM (1985) Human cytotoxic T cell clones directed against herpes simplex virus infected cells. III. Analysis of viral glycoproteins recognised by CTL clones by using recombinant herpes simplex viruses. J Immunol 134: 2679–2682

Yasukawa M et al. (1989) Differential in vitro activation of $CD4^+CD8^-$ and $CD8^+CD4^-$ herpes simplex virus-specific human cytotoxic T cells. J Immunol 143: 2051–2057

Zarling JM et al. (1986) Human cytotoxic T cell clones directed against herpes simplex virus-infected cells. IV. Recognition and activation by cloned glycoproteins gB and gD. J Immunol 136: 4669–4673

Zarling JM et al. (1988) Herpes simplex virus (HSV)-specific proliferative and cytotoxic T cell responses in humans immunised with an HSV type 2 glycoprotein subunit vaccine. J Virol 62: 4481–4485.

Cytotoxic T Lymphocyte Responses Against Measles Virus

F. G. C. M. UytdeHaag, R. S. van Binnendijk, M. J. H. Kenter, and A. D. M. E. Osterhaus

1 Introduction

In a great number of viral diseases, an important function has been attributed to cytotoxic T cells (CTL) in recovery from infection, by clearing virus-infected cells and in the induction of protective immunity against these diseases by vaccination (Yap et al. 1978; Askonas et al. 1988; Lehmann-Grube et al. 1988; Klavinskis et al. 1990; Kast et al. 1986; Reddehase et al. 1987; Cannon et al. 1987; Doherty et al. 1992). A host of questions still remains concerning the role of CTL in protection against and pathogenesis of measles. Are CTL with different phenotypes equally involved in the clearance of measles virus (MV) from infected tissue? Which MV poly- peptides and, more specifically, which epitopes of these proteins are presented by the different major histocompatibility complex (MHC) allomorphs and are recognized by major populations of CTL? Do memory CTL play a role in preventing disease upon reinfection with measles virus? Are CTL in any way

Department of Virology, Erasmus University Rotterdam, P.O. Box 1738, 3000 DR Rotterdam, The Netherlands involved in the abnormalities which may accompany acute and chronic MV infection?

Analysis of the immune response in natural MV infections and following immunization using modern tools of immunology is required to increase our understanding of the role of CTL in measles and thereby optimize the development of new generations of measles vaccines. In this chapter, we will review the present knowledge of the functions and specificities of CTL against MV.

2 Major Histocompatibility Complex Class I- and Class II-Restricted Cytotoxic T Cells

T cells recognize antigen on the surface of an antigen-presenting cell by means of a specific T cell antigen receptor (TCR). Two types of TCR exist: the $TCR_{\alpha\beta}$ and the $TCR_{\gamma\delta}$ (CLEVERS et al. 1988). T cells expressing $TCR_{\alpha\beta}$ recognize antigens (peptides) bound in the groove of MHC molecules: in humans HLA and in mice H-2 (TOWNSEND et al. 1986; BJORKMAN et al. 1987; VAN BLEEK and NATHENSON 1990; RUDENSKY et al. 1991; MONACO et al. 1992; RÖTZSCHKE et al. 1990; JORGENSEN et al. 1992). Two phenotypes of $TCR_{\alpha\beta}$-expressing T cells are distinguished, each expressing a unique cell surface glycoprotein, termed CD8 and CD4, that function as coreceptors binding to nonpolymorphic portions of the MHC antigen-presenting molecules (DOYLE and STROMINGER 1987; CONNOLLY et al. 1990). T cells expressing CD8 (T CD8) recognize antigen presented by MHC class I molecules, whereas T cells expressing CD4 (T CD4) recognize antigen presented by MHC class II molecules. Cross-linking of either CD8 or CD4 with the $TCR_{\alpha\beta}$ by the same MHC molecule is essential for T cells to become functional: T CD8 as cytotoxic T cells and T CD4 as T-helper cells (EMMRICH et al. 1986; GABERT et al. 1987; JANEWAY 1992). However, activated T CD4 can also have cytotoxic potential. In relation to possible differences between the roles that T CD8 and T CD4 play as CTLs in measles-associated immunity and pathology, it is noteworthy that class I molecules are expressed on almost all cells, while class II molecules are only expressed on cells of the immune system, notably B cells, dendritic cells, and macrophages.

The MHC class I antigen-presenting molecule is a heterodimer consisting of a polymorphic α-chain and a nonpolymorphic β_2m protein. In an exocytic vesicle, possibly the endoplasmic reticulum (ER), peptides of intracellular proteins (e.g., structural proteins of MV) may form stable complexes with the MHC class I α- and β_2m-chains, which can then be transported via the constitutive secretory pathway to the cell surface for recognition by T CD8 (YEWDELL and BENNINK 1992). MHC class II antigen-presenting molecules also exist of two polypeptide chains α and β, which become associated in ER with a third protein, the invariant chain, thereby preventing peptide binding to class II molecules in ER (NEEFJES et al. 1991). In contrast to class I, class II molecules sample peptides from degraded extracellular proteins internalized by endocytosis in an endosomal compartment (NEEFJES and PLOEGH 1992). Stable peptide–class II complexes are then transported to the cell surface for recognition by T CD4. Analysis of peptides isolated from cell lysates and from immunoaffinity purified MHC class I and class II molecules, as well as the identification of the three-dimensional structures of both types of molecules, has provided compelling evidence for an antigen-binding groove on MHC molecules that can accommodate peptides of restricted length: eight to 11 residues for class I molecules and of variable size—12–24 residues— for class II molecules (VAN BLEEK and NATHENSON 1991; FALK et al. 1991; FREMONT et al. 1992; ZHANG et al. 1992; MADDEN et al. 1991; JARDETZKY et al. 1991; HUNT et al. 1992; STERN and WILEY 1992; BJORKMAN et al. 1987; CHICZ et al. 1992; BROWN et al. 1993). In addition, these studies have initiated the analysis of peptides from a large number of antigenic proteins for residues that are critical in binding to MHC molecules or the recognition by TCR. Although many of the details of antigen processing remain to be clarified, it is clear that different MHC class I and class II molecules acquire antigenic peptides from different sources and present them to phenotypically distinct CTL. Current knowledge of antigen processing and presentation and the controversial issues linked to this subject are outlined in several excellent recent reviews (YEWDELL and BENNINK 1992; MONACO 1992; LONG 1992; BRACIALE and BRACIALE 1991; BIJLMAKERS and PLOEGH 1993).

3 Measles Virus Morphogenesis and Structure

MV, a member of the genus *Morbillivirus* of the family Paramyxoviridae, has a single negative-strand RNA genome which encodes the six structural proteins. The nucleoprotein (N), which is part of the transcription complex, forms together with the phosphoprotein (P) and the large (L) protein, which are components of the viral replication complex, the nucleocapsid containing the virus genome. The matrix (M) protein and two transmembrane glycoproteins, the hemagglutinin (H) protein and the fusion (F) protein, incorporated into a lipid bilayer contribute to the formation of the virus envelope surrounding the nucleocapsid (NORRBY and OXMAN 1990; NORRBY 1992). The H and the F transmembrane glycoproteins are cotranslationally inserted in ER of a virus-infected cell. Following glycosylation, both glycoproteins are transported via the constitutive secretory pathway to the cell surface, where they anchor in the cell membrane. The mature form of the H protein is composed of two copies of the protein forming a peplomer protruding from the virus envelope. The mature F protein consists of a nonglycosylated F1 subunit complexed to a glycosylated F2 subunit via disulphide bounds. At the cell surface, both proteins associate to form multimers before incorporation into virus particles (MORRISON and PORTNER 1992; NORRBY 1992; SATO et al. 1988; SCHEID and CHOPPIN 1977). MV is an antigenically rather stable virus, as are all paramyxoviruses. However, epitope variations do occur in the envelope glycoproteins (in the H protein more frequently than in the F protein) and in the internal proteins (in the N and P proteins less frequently than in the M protein; BARRETT 1991).

In principle, all of the viral proteins produced in a MV-infected cell can become target antigens for T CD8 and T CD4. However, with respect to the T CD4 response, it is far from clear how the cytosolic proteins, e.g., the M protein, get access to a lysosomal (MHC class II) compartment. Likewise, it is still debated where in the cell the transmembrane glycoproteins H and F, once cotranslationally inserted into ER, are degraded into MHC class I-presentable peptides. The antigen processing of MV proteins will be discussed in Sect. 6.

4 Cytotoxic T Cells in the Immune Response to Measles Virus

Recovery from many virus infections may depend on cell-mediated and antibody-mediated immunity. The importance of cell-mediated immunity in measles became first apparent from studies in immunodeficient children. Agammaglobulinemic patients who produce no detectable virus-specific antibodies showed the normal sequence of clinical symptoms and developed resistance to reinfection after contracting measles, whereas children with functional T cell deficiencies often suffered from fatal giant cell pneumonia (OLDING-STENKVIST and BJORNVATN 1976; BURNET 1968; MITUS et al. 1959; GOOD and ZAK 1956). The transient immunosuppression associated with measles (reviewed by MCCHESNEY and OLDSTONE 1989) has initially hampered investigation of cellular immunity against MV. Cytotoxic cells able to destroy MV infected target cells in vitro were demonstrated in peripheral blood leukocytes from patients with subacute sclerosing panencephalitis (SSPE), from MV-immune adults, and from children after live measles vaccination or during natural measles infection. This cellular cytotoxicity appeared to be antibody dependent (ADCC; PERRIN et al. 1977; KRETH and WIEGAND 1977; KRETH et al. 1979; WHITTLE et al. 1980). The first indications for the involvement of MHC-restricted CTL in measles came from a study by KRETH and coworkers (1979), showing that peripheral blood lymphocytes from children with acute measles killed virus-infected target cells with which they shared HLA antigens. At the same time, WRIGHT and LEVY (1979) described the generation of HLA class I-restricted CTL specific for MV from peripheral blood using MV-infected fibroblasts. MV-specific T CD8 were subsequently demonstrated in peripheral blood lymphocytes of healthy adult MV-seropositive individuals, predominantly in a high responder (LUCAS et al. 1982) or in expanded T cell cultures during secondary immune response in vitro (SETHI et al. 1982). Thus, these early studies already suggested that T CD8 may play a role in the immune response against measles and may persist as memory cells in immune individuals. The technology for establishing antigen-specific T cell lines and clones has permitted a more detailed analysis of immune responses to MV. T cell clones specific for MV were first generated in vitro from a patient with multiple sclerosis who was previously shown to be a high responder to MV. These clones had cytotoxic activity, bore the CD4 antigen, and recognized MV in association with HLA class II antigens (RICHERT et al. 1983, 1985, 1986; JACOBSON et al. 1984, 1985a, b). In addition to HLA class II-restricted T cells, HLA class I-restricted CTL could be clonally expanded in vitro from cerebrospinal fluid of a child with measles encephalitis (FLEISCHER and KRETH 1983). Subsequently, it was reported that T CD4 can be detected upon in vitro stimulation of peripheral blood lymphocytes in the majority of MV-immune subjects (JACOBSON et al. 1985b). The authors speculated that in vivo MV-specific CTL are predominantly T CD4 (JACOBSON et al. 1987/1988). This was, however, not fully in agreement with the earlier findings on T CD8 in MV-immune individuals. At present, it is clear that the antigen presentation form (infectious virus versus noninfectious particles or soluble viral proteins) influences in vitro induction and frequency of T CD4 and T CD8 in different viral systems (MORRISON et al. 1988; VAN BINNENDIJK et al. 1989, 1992a). Thus, in contrast to earlier reports (McFARLAND et al. 1980; GREENSTEIN and McFARLAND 1983), it was shown that peripheral blood mononuclear cells respond upon in vitro stimulation with MV antigen in the majority (75%) of measles immune adults. Depending on the form of antigen presentation—Epstein-Barr virus-transformed B cell lines (B-LCL) either infected with live virus or pulsed with UV-inactivated virus—we could isolate T CD8 and T CD4 clones from such cultures, suggesting the existence of MV-specific memory at both the T CD4 and T CD8 level (VAN BINNENDIJK et al. 1989). Recently, using synthetic nonamer peptides representing HLA-A2.1-presentable epitopes of MV, which do not require additional processing by antigen-presenting cells, KRETH and coworkers showed the generation of MV-specific T CD8 from peripheral blood of MV-immune individuals (NANAN et al. 1992). In addition, we identified and cloned T CD8 as effector cells predominating in MV-specific immune responses in three children with acute measles (VAN BINNENDIJK et al. 1990, 1991). Substantiating this finding, in 31 out of 48 vaccinated schoolchildren involved in a measles outbreak, T CD8 could be expanded from peripheral blood lymphocyte cultures upon specific stimulation with live MV (VAN BINNENDIJK et al. 1992b). Investigating the pathogenesis of MV-induced encephalitis in a mouse model, NIEWIESK and associates (1993) found that T CD8 are an important component of the immune response to MV. Finally, WILD and colleagues demonstrated the induction of T CD8 in mice following immunization with vaccinia virus recombinants coding for MV proteins (BEAUVERGER et al. 1993). A number of studies have demonstrated a role for T CD4 as well as T CD8 in protection against MV-induced encephalitis (BANKAMP et al. 1991; BRINCKMANN et al. 1991; NIEWIESK et al. 1993; DE VRIES et al. 1988). It is unsolved whether in these models T CD4 participate in conferring protection by functioning as CTL. The absence of MHC class II expression in cells of the brain would render them unsusceptible for lysis by T CD4, though virus infection may induce γ-interferon production leading to expression of MHC class II molecules on the potential targets. In the study by NIEWIESK et al. (1993), T CD4 were generally only weakly lytic, and susceptibility to MV-induced encephalitis clearly correlated with inefficient induction of T CD8. Also, in rats depleted of T CD8, persistence of MV in brain neurons is promoted (MAEHLEN et al. 1989). It has been suggested that T CD4 with cytolytic potential may terminate ongoing immune responses by killing antigen-presenting cells rather than eliminate infected target cells early in infection (BRAAKMAN et al. 1987). Clearly, additional studies are necessary to determine the relative roles of both T CD4 and T CD8 in protection against measles in vivo, the outcome of which will aid to optimize the design of new measles vaccines (see Sect. 7).

Collectively, the recent data indicate that, as is the case in other virus infections (ASKONAS et al. 1988; LEHMANN-GRUGE et al. 1988), T CD8 are involved in recovery from measles and may function as memory cells in maintaining lifelong immunity against measles.

5 Epitopes of Measles Virus Proteins for T CD4 and T CD8

Further analysis of antigenic structures and of the role of different structural proteins and epitopes thereof in the induction of protective immunity is likely to be important in the development of new generations of vaccines against measles. It may be speculated that the induction of specific CTL is an important aspect of this immunity. It has been shown that in principle all the structural proteins of MV can elicit specific T cell reactivity in human individuals, as assayed by proliferative responses of lymphocytes in bulk cultures or of T cell lines or clones to purified proteins of MV (BELLINI et al. 1981; ROSE et al. 1984; RICHERT et al. 1986; JACOBSON et al. 1987/1988, 1989; DHIB-JALBUT et al. 1988; VAN BINNENDIJK et al. 1989; ILONEN et al. 1990). Here we discuss the reported T cell epitopes of MV proteins as found in man, mice, and rats.

5.1 Fusion Protein

Identification of MHC class I- and class II-restricted T cell epitopes of the F protein of MV has been subject of extensive studies. Using genetically engineered F protein, nested sets of synthetic peptides spanning the F protein and T cell clones, a number of antigenic regions for T CD4 and T CD8 were identified. The core sequences required for HLA binding and T cell recognition were determined by using combined N- and C-terminal deleted peptides (VAN BINNENDIJK et al. 1993; see Fig. 1). Three T CD4 epitopes, an HLA DR2- (aa 379–390), an HLA DQ 5,6- (aa 427–435), and an HLA DRw53- (aa 454–463) restricted epitope were identified that cluster between aa 379–466. One peptide recognized by T CD4 in association with HLA DQw1 also

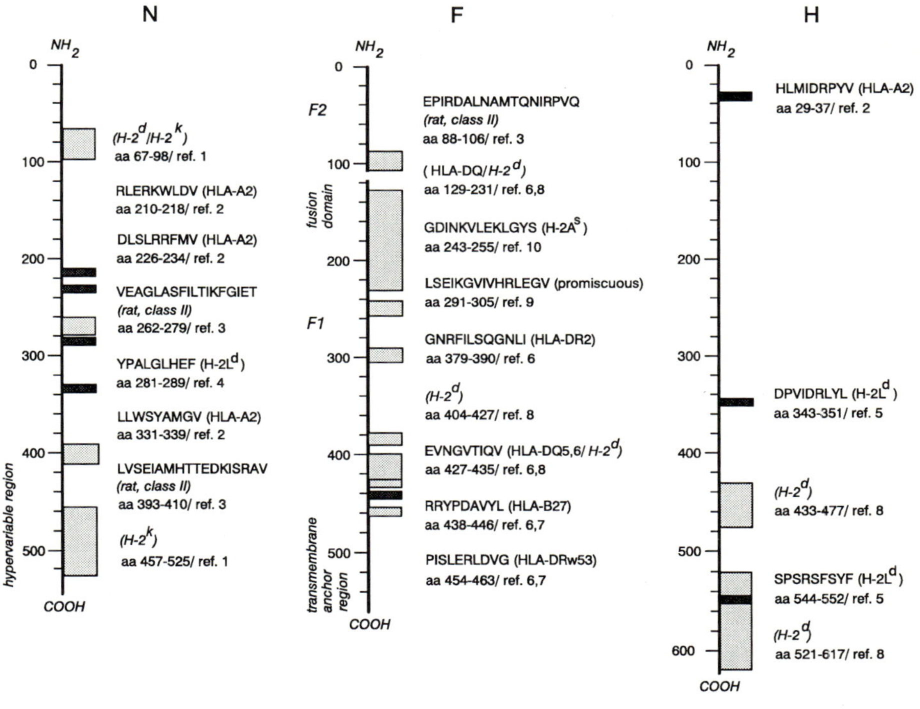

Fig. 1. T cell epitopes identified on the nucleoprotein (*N*), fusion protein (*F*), and hemagglutinin protein (H) of measles virus. Sequence of the N protein according to BUCKLAND et al. (1988); the H protein according to ALKHATIB and BRIEDIS (1986); and the F protein according to RICHARDSON et al. (1986), with the numbering of amino acid sequences starting at the first methionine residue. *ref. 1*, GIRAUDON et al. (1991); *ref. 2*, NANAN et al. (1992); *ref. 3*, Liebert (personal communication); *ref. 4*, BEAUVERGER et al. (1993); *ref. 5*, Beauverger (personal communication); *ref. 6*, VAN BINNENDIJK et al. (1993); *ref. 7*, VAN BINNENDIJK et al. (1992a); *ref. 8*, Versteeg (personal communication); *ref. 9*, PARTIDOS and STEWARD (1990); PARTIDOS et al. (1991); *ref. 10*, PARTIDOS and STEWARD (1992). *Dark bars* represent epitopes recognized by T CD8; *light bars* represent epitopes recognized by T CD4

represents a T cell epitope for mice of the H-2^d haplotype (VAN BINNENDIJK et al. 1993; Versteeg, personal communication). An additional mouse (H-2^d) class II-restricted epitope, from which the minimal sequence has not yet been determined, is located between aa 404–427 of the F protein (Versteeg, personal communication). In the same region of the F protein, a nonamer sequence (aa 438–446) was found that matches the HLA B27 binding motif. The synthetic peptide corresponding to aa 438–446 indeed sensitized target cells for recognition by the HLA B27-restricted T CD8 clone. The presently known T CD4 epitopes of the F protein all have the common features of MHC class II binders (VAN BINNENDIJK et al. 1993). Using several criteria that predict sequence motifs binding to MHC class II molecules, PARTIDOS et al. (1991; PARTIDOS and STEWARD 1990, 1992) identified two sequences on the F protein that indeed appeared to represent T cell epitopes (Fig. 1). The region aa 291–305 contains an immunodominant epitope of promiscuous nature, since the corresponding peptide induced MV-specific proliferative T cell responses in different haplotype strains of mice and proliferative responses in peripheral blood lymphocytes of apparently HLA- disparate MV-immune individuals. The other region, aa 243–255, contains an H-$2A^s$-restricted T cell epitope. The region aa 129–231 contains an HLA DQ-restricted epitope recognized by cloned T CD4 which also appears to be a class II-restricted epitope for mice of the H-2^d haplotype (VAN BINNENDIJK et al. 1993 and unpublished results). Recently, an epitope has been identified at position aa 86–106 of the F protein. A synthetic peptide corresponding to this sequence stimulates the in vitro proliferation of lymphocytes from MV-immune rats (Liebert, personal communication).

5.2 Nucleoprotein

Several T CD4 and T CD8 epitopes have been identified on the N protein of MV. In a data base search NANAN et al. (1992 and personal communication) screened the N protein for epitopes matching the anchor sequence of HLA A2. On basis of this search, three nonamer peptides corresponding to the N protein sequences aa 210–218, 226–234, and 331–339 were synthesized that were shown to induce MV-specific T CD8 from peripheral blood lymphocytes of HLA A2-positive MV-immune individuals (Fig. 1). An epitope recognized by mouse T CD8 in association with H-$2L^d$ has been identified on the N protein (aa 281–289) by scanning the sequence of the protein for H-$2L^d$ motifs and testing the corresponding peptides for their capacity to sensitize target cells for killing by MV-specific T CD8 (BEAUVERGER et al. 1993; Fig. 1). This epitope appeared also to be conserved in the N protein of canine distemper virus, a morbillivirus closely related to MV. Two regions on the N protein, possibly containing H-2 class II-restricted antigenic epitopes for T-helper cells, were reported (GIRAUDON et al. 1991). An immunodominant epitope is located in a highly conserved region near the N terminus (aa 67–98), whereas a second epitope which is antigenic in mice of the H-2^k haplotype is located in the hypervariable region near the C terminus of the N protein (Fig. 1). Finally, two class II-restricted epitopes for rat T cells have been identified recently, using synthetic peptides corresponding to the aa 262–279 and aa 393–410 sequences of the N protein (Liebert, personal communication).

5.3 Hemagglutinin Protein

The mapping of T cell epitopes on the H protein has started only recently. Thus, an HLA A2-restricted epitope has been identified near the N terminus on the H protein (aa 29–37) which is recognized by T CD8 (Nanan et al. 1992 and personal communication). Using polyclonal H protein-reactive CTL generated in Balb/c (H-2^d) mice and nonamer peptides corresponding to different H-$2L^d$ motifs present in the primary sequence of the H protein, two T cell epitopes on the H protein were identified. With two peptides corresponding to aa 343–351 and aa 544–552, respectively, similar levels of target cell lysis were found as with targets that constitutively expressed the H protein (Beauverger, personal communication; Fig. 1). Finally, using genetically engineered H proteins and class II-restricted cloned T cells we have identified two regions (aa 433–477 and aa 521–617) containing putative T cell epitopes for mice of the H-2^d haplotype (Versteeg, personal communication; Fig. 1). It is expected that in the near future more epitopes will be identified on these important structural proteins of MV.

5.4 Large and Matrix Proteins

Thus far only one, HLA A2-restricted, epitope for T CD8 has been identified on the L protein (Nanan et al. 1992). Based on recently determined HLA DQw2 binding motif, a putative epitope can be identified on the M protein of MV, and a peptide corresponding to this sequence is at presently being tested for binding to HLA DQW2 (Koning, personal communication).

In conclusion, it appears that T cell epitopes of the F, H, and N proteins of MV are scattered along the entire sequence of these structural proteins and that different MHC class I and class II molecules bind different sets of peptides. Size of peptides and conservation of "anchor" residues in peptide sequences are shown to be critical for allele-specific binding. Due to these allele-specific peptide motifs, it has been possible to predict allele-specific binding motifs in the protein sequences of the structural proteins of MV. Testing of the corresponding peptides for binding to the respective alleles and for their capacity to elicit T cell responses in vitro using lymphocytes of MV-immune individuals has identified T cell epitopes on MV proteins. The binding affinity of a MV peptide for an MHC molecule and subsequently the affinity of the TCR for the peptide–MHC complex largely determines the immunogenicity of an epitope. Thus, the MHC class I and class II haplotype differences among individuals in the human outbred population may be responsible for variations in the immune response of different individuals to the same antigens. The identification of MV peptide sequences which correspond with binding motifs of the most frequent MHC allomorphs may lead to the characterization of immunodominant T cell epitopes of MV proteins and is therefore important for the development of new generations of MV vaccines.

6 Processing of Cytosolic and Transmembrane Proteins of Measles Virus

As briefly mentioned in Sect. 2, it is now widely appreciated that MHC class I and class II molecules travel along different biosynthetic routes to the cell surface and sample different cellular compartments for antigenic peptides (YEWDELL and BENNINK 1992; NEEFJES and PLOEGH 1992). All data thus far available are consistent with the notion that MHC class I molecules are loaded with peptides in ER or *cis*-Golgi complex and with a role for the TAP1–TAP2 peptide transporter complex in the import of peptides from the cytosol to these compartments (BRACIALE and BRACIALE 1991; MONACO 1992; BIJLMAKERS and PLOEGH 1993). In addition, the generation of antigenic peptides for presentation by MHC class I molecules is believed to take place in the cytosol. While appealing for presentation of peptides originating from proteins biosynthesized in the cytosol, this general view of antigen presentation seems difficult to reconcile with presentation of transmembrane proteins by MHC class I molecules. Thus, in MV-infected cells the H and F transmembrane proteins are cotranslationally imported in ER and yet MHC class I molecules present peptides of these proteins at the cell surface (VAN BINNENDIJK et al. 1992a). Are these peptides generated in ER? Proteolytic degradation can indeed occur in ER, but is rather selective (BONIFACINO et al. 1990), and apart from generation of signal sequence-derived peptides, there is no evidence supporting generation of MHC class I-presentable peptides of transmembrane proteins in ER. Is it possible that peptides for MHC class I presentation originate from cytosolic processing of transmembrane proteins and, if so, how do these proteins reach the cytosol? We have recently addressed these questions using antigen presentation-defective mutant cell lines and a cloned T CD8 recognizing the nonamer peptide sequence RRYPDAVYL of the F protein of MV in association with HLA B27 (Fig. 1). The mutant T2 cell line transfected with the HLA B27 gene but lacking high expression of HLA B2705 molecules at the cell surface, when infected with MV, did not function as a target cell for lysis by the T CD8 clone (VAN BINNENDIJK et al. 1992a). As T2 cells lack multiple genes that may be involved in antigen presentation, including *TAP1* and *TAP2* genes and the *LMP-2* and *LMP-7* genes encoding the proteasome components (DE MARS et al. 1984; SALTER and CRESSWELL 1986; KELLY et al. 1991; GLYNNE et al. 1991; YANG et al. 1992), these results suggest that at least one gene, which is absent in T2, plays a role in presentation of this epitope of the F protein by MHC class I molecules. Using the TAP2 peptide transporter deficient EBV transformed B cell line BM 36.1 expressing HLA B2705, we recently obtained evidence that the presentation of the HLA B27-restricted epitope of the F protein of MV by an EBV transformed B cell line requires a functional TAP1–TAP2 peptide transporter (Kenter et al. submitted). Thus, it appears that the cytosol is a processing compartment yielding MHC class I-presentable peptides of ER luminal domains of transmembrane proteins. How these transmembrane proteins get access to the cytosol is as yet unknown.

In contrast to MHC class I molecules, MHC class II molecules sample endocytic compartments for antigenic peptides (NEEFJES et al. 1991). However, there is evidence from several studies employing different experimental systems, that epitopes of endogenous cytosolic proteins are presented by MHC class II molecules (see LONG 1992 for review). In principal, several mechanisms could explain presentation of cytosolic proteins by MHC class II molecules. Most unlikely, cytosolic processing may yield peptides loading MHC class II molecules in ER even in the presence of the invariant chain (NEWCOMB and CRESSWELL 1993). Secondly, autophagy, sequestration of cytosolic material by ER derived membranes, may result in the delivery of cytosolic proteins to autophagic lysosomes. This mechanism could perhaps also explain those examples of MHC class II-mediated presentation of transmembrane proteins that are resistant to acidotropic agents such as chloroquine, primaquine, and ammonium chloride (VAN BINNENDIJK et al. 1992a). Finally, chaperoning molecules may be involved in the transport of cytosolic proteins or peptides thereof to compartments that are sampled by class MHC II molecules (see also review by LONG 1992).

With the rapid progress in understanding antigen presentation by MHC class I molecules to T CD8, development of new vaccination strategies has coevolved. In particular the design of vehicles to import exogenous protein antigens into the class I antigen presentation route has received considerable attention.

7 Alternative Strategies for Vaccination Against Measles: A Focus on Cytotoxic T Cells

Only a few years after MV had been adapted to in vitro cell cultures systems, the first live attenuated measles vaccines based on the Edmonston B strain of MV were prepared (KATZ 1965). Although live measles vaccines have been highly successful, there are still considerable disadvantages associated with their use. Perhaps the most important of these is that they cannot be used in the presence of maternally derived MV-specific antibodies, which is a major drawback for their use in developing countries. Almost simultaneously with the introduction of live attenuated measles vaccines, Tween-ether- and formaldehyde-inactivated whole virus preparations were introduced as measles vaccines. Although the majority of these preparations were shown to induce MV-neutralizing and hemagglutination-inhibiting serum antibodies, these antibodies were shown to persist only for relatively short periodes, compared to antibodies induced by natural MV infection or vaccination with live attenuated measles vaccines. More important was the observation that individuals vaccinated with the inactivated vaccines were often not protected against MV infection and that their symptoms were often more severe than those observed in nonvaccinated children. This manifestation of measles became known as atypical measles syndrome (FULGINITI and HELFER 1980). NORRBY and coworkers (1975) suggested that an inappropriate immune response to the F protein of the virus, which resulted in the absence of biologically active fusion-inhibiting antibodies, was at the basis of this failure. As we and others have shown, MV-specific T CD8 may play an important role in recovery from measles (see Sect. 4) and it may not be expected that the inactivated measles vaccines used would have specifically stimulated this population of T lymphocytes; the absence of this T cell response may, therefore, be another or additional explanation for the failure of this type of vaccines. Consequently, an important requirement for novel generations of measles vaccines should be that they elicit a MV-specific T CD8 response and memory and this preferably in the presence of MV-specific maternally derived antibodies. Evaluating the present candidates for novel generations of measles vaccines in this light, the two most promising approaches seem the use of live recombinant viruses or bacteria and immune-stimulating complex the (ISCOM) matrix for the presentation of the MV glycoproteins (for review see OSTERHAUS and DE VRIES 1992). It may be expected that live recombinant vaccinia and avipox viruses, expressing the F and the H protein of MV, which are considered serious candidates for measles vaccines, will indeed induce the required T CD8 response and memory, since de novo synthesis of the respective proteins is established after immunization. If administered along the proper routes, they may also be effective in the presence of maternally derived MV-specific serum antibodies. Although nonreplicating subunit vaccines are generally not capable of inducing a specific T CD8 response, we and others have recently shown that when incorporated in the ISCOM matrix, viral and other proteins may elicit a specific T CD8 response in vivo and in vitro (for review see CLAASSEN and OSTERHAUS 1992). As mentioned in Sect. 6, this is probably related to the observation that proteins presented by the ISCOM matrix, are processed for HLA class I-restricted CTL recognition in a way similar to de novo synthesized proteins. We showed that live MV and MV-F ISCOM allow presentation by MHC class I molecules, while other nonreplicating presentations of the F protein only permitted MHC class II-restricted presentation (VAN BINNENDIJK et al. 1992a). This observation, together with the demonstration that F-ISCOM are potent inducers of long-lasting specific serum antibodies (even in the presence of specific serum antibodies), which include biologically active F-specific antibodies and of solid protection against morbillivirus infections (for review see OSTERHAUS and DE VRIES 1992), indicates that the ISCOM presentation form is a serious candidate for the development of a new generation of measles vaccines.

Acknowledgements. The authors wish to thank Dr. W. Kreth, Dr. R. Nanan, Dr. U. Liebert, Dr. F. Wild, Dr. P. Beauverger, Dr. J. Versteeg, Dr. F. Koning, and Dr. J. H. Brown for kindly sharing their unpublished data and Conny Kruyssen for help preparing this manuscript.

References

Alkhatib G, Briedis DJ (1986) The predicted primary structure of the measles virus hemagglutinin. Virology 150: 479–484

Askonas BA, Taylor PM, Esquivel F (1988) Cytotoxic T cells in influenza infection. Ann N Y Acad Sci 532: 230–237

Bankamp B, Brinckmann UG, Reich A, Niewiesk S, ter Meulen V, Liebert UG (1991) Measles virus nucleocapsid protein protects rats from encephalitis. J Virol 65: 1695–1700

Barrett T, Subbaroa SM, Belsham GJ, Mahy BW (1991) The molecular biology of morbilliviruses. In: Kingsbury DW (ed) The paramyxoviruses. Plenum, New York, pp 83–102

Beauverger P, Buckland R, Wild TF (1993) Measles virus antigen induces both type specific and canine distemper cross reactive CTLs in mice: localisation of a common NP L^d-restricted epitope. J Gen Virol (in press)

Bellini WJ, McFarlin DE, Silver GD, Mingioli ES, McFarland HF (1981) Immune reactivity of the purified hemagglutinin of measles virus. Infect Immun 32: 1051–1057

Bijlmakers MJ, Ploegh HL (1993) Putting together an MHC class I molecule. Curr Biol 5: 21–26

Bjorkman PJ, Saper MA, Samraoui B, Bennet WS, Strominger JL, Wiley DC (1987) The foreign antigen binding site and T cell recognition regions of class I histocompatibility antigens. Nature 329: 512–518

Bonifacino JS, Cosson P, Klausner RD (1990) Colocalized transmembrane determinants for ER degradation and subunit assembly explain the intracellular fate of TCR chains. Cell 63: 503–513

Braakman E, Rotteveel FTM, van Bleek G, van Seventer GA, Lucas K (1987) Are MHC class II-restricted cytotoxic T lymphocytes important? Immunol Today 8: 265–267

Braciale TJ, Braciale VI (1991) Antigen presentation: structural themes and functional variations. Immunol Today 12: 124–129

Brinckmann UG, Bankamp B, Reich A, ter Meulen V, Liebert AG (1991) Efficacy of individual measles virus structural proteins in the protection of rats from measles encephalitis. J Gen Virol 72: 2491–2500

Brown JH, Jardetzky TS, Gorga JC, Stern LJ, Urban RG, Strominger JL, Wiley DC (1993) The three-dimensional structure of the human class II histocompatibility antigen HLA-DR1. Nature (in press)

Buckland R, Gerald C, Barker D, Wild TF (1988) Cloning and sequencing of the nucleoprotein gene of measles virus (Hallé strain). Nucleic Acids Res 16: 11821–11826

Burnet FM (1968) Measles as an index of immunological function. Lancet 2: 610–613

Cannon J, Scott EJ, Taylor G, Askonas BA (1987) Clearance of persistent respiratory syncytial virus infection in immunodeficient mice following transfer of primed T cells. Virology 62: 133–183

Chicz RM, Urban RG, Lane WS, Gorga JC, Stern LJ, Vignali DAA, Strominger JL (1992) Predominant naturally processed peptides bound to HLA-DR1 are derived from MHC-related molecules and are heterogeneous in size. Nature 358: 764–768

Claassen I, Osterhaus A (1992) The iscom structure as an immune enhancing moiety: experience with viral systems. Res Immunol 143 (5): 531–541

Clevers H, Alarcon B, Wileman T, Terhorst C (1988) The T cell receptor/CD3 complex: a dynamic protein ensemble. Annu Rev Immunol 6: 629–662

Connally JM, Hansen TH, Ingold HL, Potter TA (1990) Recognition by CD8 by cytotoxic T lymphocytes is ablated by several subtitutions in the class I $\alpha 3$ domain: CD8 and the T cell receptor recognize the same class I molecule. Proc Natl Acad Sci USA 87: 2137–2141

DeMars R, Chang CC, Shaw S, Reitnauer PJ, Sondel PM (1984) Homozygous deletion that simultaneously eliminates expression of class I and class II antigens of EBV-transformed B-lymphoblastoid cells. Hum Immunol 11: 77–80

De Vries P, van Binnendijk RS, van der Marel P, van Wezel AL, Voorma HO, Sundquist B, UytdeHaag FGCM, Osterhaus ADME (1988) Measles virus fusion protein presented in immune-stimulating complex (ISCOM) induces haemolysis-inhibiting and fusion-inhibiting antibodies, virus-specific T cells and protection in mice. J Gen Virol 69: 549–559

Dhib-Jalbut S, McFarland HF, Mingioli ES, Sever JL, McFarlin (1988) Humoral and cellular immune responses to matrix protein of measles virus in subacute sclerosing panencephalitis. J Virol 62: 2483–2489

Doherty PC, Allan W, Eichelberg M (1992) Roles of $\alpha\beta$ and $\tau\delta$ T cell subsets in viral immunity. Annu Rev Immunol 10: 123–151

Doyle C, Strominger JL (1987) Interaction between CD4 and class II MHC molecules mediates cell adhesions. Nature 330: 256–258

Emmrich F, Strittmatter U, Eichmann K (1986) Synergism in the activation of human CD8 T cells by crosslinking the T cell receptor complex with CD8 differentiation antigen. Proc Natl Acad Sci USA 83: 8292–8302

Falk K, Rötschke O, Stevanovic S, Jung G, Rammensee H-G (1991) Allele specific motifs revealed by sequencing of self peptides eluted from MHC molecules. Nature 351: 290–296

Fleischer B, Kreth HW (1983) Clonal expansion and functional analysis of virus-specific T lymphocytes from cerebrospinal fluid in measles encephalitis. Hum Immunol 7: 239–248

Fremont DH, Matsummura M, Stura EA, Peterson PA, Wilson IA (1992) Crystal structures of two viral peptides in complex with murine H-2K^b. Science 257: 991–927

Fulginiti VA, Helfer RE (1980) Atypical measles in adolescent siblings 16 years after killed measles virus vaccine. J Am Med Assoc 244: 804–806

Gabert J, Langlet C, Zamoyska R, Parnes JR, Schmitt-Verhulst AM, Malissen B (1987) Reconstitution of MHC class I specificity by transfer of the T cell receptor and Lyt-2 genes. Cell 50: 545–554

Giraudon P, Buckland R, Wild TF (1991) The immune response to measles virus in mice. T-helper response to the nucleoprotein and mapping of the T-helper epitopes. Virus Res 22: 41–54

Glynne R, Powis SH, Beck S, Kelly A, Kerr LA, Trowsdale J (1991) A proteasome-related gene between the two ABC transporter loci in the class II region of the human MHC. Nature 353: 357–360

Good RA, Zak SJ (1956) Disturbances in gamma globulin synthesis as 'experiments of nature.' Paediatrics 18: 109–149

Greenstein JI, McFarland HF (1983) Response of human lymphocytes to measles virus after natural infection. Infect Immun 40: 198–204

Hunt DF, Michel H, Dickinson T, Shabanowitz J, Cox AL, Sakaguchi E, Appella A, Grey HM, Sette A (1992) Peptides presented to the immune system by the murine class II MHC molecules I-Ad. Science 256: 1817–1820

Ilonen J, Mäkelä MJ, Ziola B, Salmi A (1990) Cloning of human T cells specific for measles virus haemagglutinin and nucleocapsid. Clin Exp Immunol 81: 212–217

Jacobson S, Richert JR, Biddison WE, Satinsky A, Hartzman RJ, McFarland HF (1984) Measles virus-specific T4 human cytotoxic T cell clones are restricted by class II HLA antigens. J Immunol. 33: 754–757

Jacobson S, Flerlage ML, McFarland HF (1985a) Impaired measles virus-specific cytotoxic T cell responses in multiple sclerosis. J Exp Med 162: 839–850

Jacobson S, Nepom GT, Richert JR, Biddison WE, and McFarland HF (1985b) Identification of a specific HLA DR2 Ia molecule as a restriction element for measles virus-specific HLA class II-restricted cytotoxic T cell recognition. J Exp Med 161: 263–268

Jacobson S, Rose JW, Flerlage ML, McFarlin DE, McFarland HF (1987/1988) Induction of measles virus-specific human cytotoxic T cells by purified measles virus nucleocapsid and hemagglutinin polypeptides. Viral Immunol 1: 153–162

Jacobson S, Sekaly RP, Jacobson CL, McFarland HF, Long EO (1989) HLA class II-restricted presentation of cytoplasmic measles virus antigens to cytotoxic T cells. J Virol 63: 1756–1762

Janeway CA Jr (1992) The T cell receptor as a multicomponent signalling machine: CD4/CD8 coreceptors and CD45 in T cell activation. Annu Rev Immunol 10: 645–674

Jardetzky TS, Lane WS, Robinson RA, Madden DR, Wiley DC (1991) Identification of self peptides bound to purified HLA-B27. Nature 353: 326–329

Jorgensen JL, Reay PA, Ehrich EW, Davis MM (1992) Molecular components of T cell recognition. Annu Rev Immunol 10: 835–873

Kast WM, Bronkhorst AM, De Waal LP, Melief CJM (1986) Cooperation between cytotoxic and helper T lymphocytes in protection against a lethal sendai virus infection. J Exp Med 164: 723–738

Katz SL (1965) Immunization with live attenuated measles virus vaccines: five years experience. Arch Gesamte Virusforsch 16: 222–230

Kelly A, Powis S, Glynee R, Radley E, Beck S, Trowsdale J (1991) Second proteasome-related gene in the human MHC class II region. Nature 353: 667–668

Klavinskis LS, Whitton JL, Joly E, Oldstone MBA (1990) Vaccination and protection from a lethal virus infection: identification, incorporation and use of a cytotoxic T lymphocyte glycoprotein epitope. Virology 178: 393–400

Kreth HW, Wiegand G (1977) Cell-mediated cytotoxicity against measles virus in SSPE II. Analysis of cytotoxic effector cells. J Immunol 118: 296–302

Kreth HW, ter Meulen V, Eckert G (1979) Demonstration of HLA restricted killer cells in patients with acute measles. Med Microbiol Immunol 165: 203–214

Lehmann-Grube FD, Moskophidis D, Lohler J (1988) Recovery from acute virus infection. Ann N Y Acad Sci 532: 238–255

Long EO (1992) Antigen processing for presentation to $CD4^+$ T cells. New Biologist 4: 274–282

Lucas CJ, Biddison WE, Nelson DL, Shaw S (1982) Killing of measles virus-infected cells by human cytotoxic T cells. Infect Immun 38: 226–232

Madden DR, Gorga JC, Strominger JL, Wiley DC (1991) The structure of HLA-B27 reveals nonamer self peptides bound in an extended conformation. Nature 353: 321–325

Maehlen J, Olsson T, Löve A, Klareskog L, Norrby E (1989) Persistence of measles virus in rat brain neurons is promoted by depletion of $CD8^+$ T cells. J Neuroimmunol 21: 149–155

McChesney MB, Oldstone MBA (1989) Virus-induced immunosuppression: infections with measles virus and human immunodeficiency virus. Adv Immunol 45: 355–380

McFarland HF, Pedone CA, Mingioli ES, McFarlin DE (1980) The response of human lymphocyte subpopulations to measles, mumps and vaccinia viral antigens. J Immunol 125: 221–225

Mitus A, Enders JF, Craig JM, Holloway A (1959) Persistence of measles virus and depression of antibody formation in patients with giant cell pneumonia after measles. N Engl J Med 261: 882–889

Monaco JJ (1992) Pathways of antigen processing. A molecular model of MHC class I restricted antigen processing. Immunol Today 13: 173–178

Morrison LA, Braciale VL, Braciale TJ (1988) Antigen form influences induction and frequency of influenza-specific class I and class II MHC-restricted cytolytic T lymphocytes. J Immunol 141: 363–368

Morrison T, Portner A (1992) Structure, function, and intracellular processing of the glycoproteins of paramyxoviridae. In: Kingsbury DW (ed) The paramyxoviruses. Plenum, New York, pp 347–382

Nanan R, Petzold K, Carstens C, Kreth HW (1992) Correct prediction of HLA-A2.1 epitopes of measles virus. Immunobiology 186: 10

Neefjes JJ, Ploegh HI (1992) Intracellular transport of MHC class II molecules. Immunol Today 5: 179–184

Neefjes JJ, Schumacher TNM, Ploegh HL (1991) Assembly and intracellular transport of major histocompatibility complex molecules. Curr Opin Cell Biol 3: 601–609

Newçomb JR, Cresswell P (1993) Characterization of endogenous peptides bound to purified HLA-DR molecules and their absence from invariant chain-associated $\alpha\beta$ dimers. J Immunobiol 150: 499–507

Niewiesk S, Brinckmann U, Bankamp B, Sirak S, Liebert UG, ter Meulen V (1993) Susceptibility to measles virus-induced encephalitis in mice correlates with impaired antigen presentation to cytotoxic T lymphocytes. J Virol 67: 75–81
Norrby E (1992) Immunobiology of paramyxoviruses. In: Kingsbury DW (ed) The paramyxoviruses. Plenum, New York, pp 481–507
Norrby E, Oxman MN (1990) Measles virus. In: Fields BN (ed) Virology, vol 1. Raven, New York, pp 1013–1044
Norrby E, Enders-Ruckle G, ter Meulen V (1975) Differences in the appearance of antibodies to structural components of measles virus after immunization with inactivated and live virus. J Infect Dis 132: 262–269
Olding-Stenkvist E, Bjorvatn B (1976) Rapid detection of measles virus in skin rashes by immunofluorescence. J Infect Dis 134: 463–469
Osterhaus ADME, de Vries P (1992) Vaccination against acute respiratory virus infections and measles in man. Immunobiology 184: 180–192
Partidos CD, Steward MW (1990) Prediction and identification of a T cell epitope in the fusion protein of measles virus immunodominant in mice and humans. J Gen Virology 71: 2099–2105
Partidos CD, Steward MW (1992) The effects of a flanking sequence on the immune response to a B and a T cell epitope from the fusion protein of measles virus. J Gen Virol 73: 1987–1994
Partidos CD, Stanley CM, Steward MW (1991) Immune responses in mice following immunization with chimeric synthetic peptides representing B and T cell epitopes of measles virus proteins. J Gen Virol 72: 1293–1299
Perrin LH, Tishon A, Oldstone MBA (1977) Immunologic injury in measles infection. III. Presence and characterization of human cytotoxic lymphocytes. J Immunol 118: 282–290
Reddehase MJ, Mutter W, Munch K, Buhring HJ, Koszinowski UH (1987) CD8-positive T lymphocytes specific for murine cytomegalovirus immediate-early antigens mediate protective immunity. J Virol 61: 3102–3108
Richardson C, Hull D, Greer P, Hasel K, Berkovich A, Englund G, Bellini W, Rima B, Lazzarini R (1986) The nucleotide sequence of the mRNA encoding the fusion protein of measles virus (edmonston strain): a comparison of fusion proteins from several different paramyxoviruses. Virology 155: 508–523
Richert JR, McFarland HF, McFarlin DE, Bellini WJ, Lake P (1983) Cloned measles virus-specific T lymphocytes from a twin with multiple sclerosis. Proc Natl Acad Sci USA 80: 555–559
Richert JR, McFarland HF, McFarlin DE, Johnson AH, Woody JH, Hartzman RJ (1985) Measles-specific T cell clones derived from a twin with multiple sclerosis: genetic restriction studies. J Immunol 34: 1561–1566
Richert JR, Rose JW, Reuben-Burnside C, Kearns MC, Jacobson S, Mingioli ES, Hartzman RJ, McFarland HF, McFarlin DE (1986) Polypeptide specificities of measles virus-reactive T cell lines and clones derived from a patient with multiple sclerosis. J Immunol 137: 2190–2194
Rose JW, Bellini WJ, McFarlin DE, McFarland HF (1984) Human cellular immune response to measles virus polypeptides. J Virol 49: 988–991
Rötzschke O, Falk F, Deres K, Schild H, Norda M, Metzger J, Jung G, Rammensee HG (1990) Isolation and analysis of naturally processed viral peptides as recognized by cytotoxic T cells. Nature 348: 252–254
Rudensky AY, Preston-Hurlburt P, Hong S-C, Barlow A, Janeway CA Jr (1991) Sequence analysis of peptides bound to MHC class II molecules. Nature 353: 622–626
Salter RD, Cresswell P (1986) Impaired assembly and transport of HLA-A and -B antigens in a mutant TxB cell hybrid. EMBO J 5: 943–949
Sato TA, Kohama T, Sugiura A (1988) Intracellular processing of measles virus fusion protein. Arch Virol 98: 39–50
Scheid A, Choppin PW (1977) Two disulphide-linked chains constitute the active F protein of paramyxoviruses. Virology 80: 54–66
Sethi KK, Stroehmann I, Brandis H (1982) Generation of cytolytic T-cell cultures displaying measles virus specificity and human histocompatibility leukocyte antigen restriction. Infect Immun 36: 657–661
Stern LJ, Wiley DC (1992) The human class II MHC protein HLA-DR1 assembles as empty $\alpha\beta$ heterodimers in the absence of antigenic peptide. Cell 68: 465–477
Townsend ARM, Rothbard J, Gotch FM, Bahadur G, Wraith D, McMichael J (1986) The epitopes of influenza nucleoprotein recognized by cytotoxic T lymphocytes can be defined with short synthetic peptide. Cell 44: 959–968

Van Binnendijk RS, Poelen MCM, de Vries P, Voorma HO, Osterhaus ADME, UytdeHaag FGCM (1989) Measles virus-specific human T cells clones. Characterization of specificity and function of $CD4^+$ helper/cytotoxic and $CD8^+$ cytotoxic T cell clones. J Immunol 142: 2847–2854

Van Binnendijk RS, Poelen MCM, Kuijpers KC, Osterhaus ADME, UytdeHaag FGCM (1990) The predominance of $CD8^+$ T cells after infection with measles virus suggests a role for $CD8^+$ class I MHC-restricted cytotoxic T lymphocytes (CTL) in recovery from measles. Clonal analyses of human $CD8^+$ class I MHC-restricted CTL. J Immunol 144: 2394–2399

Van Binnendijk RS, Poelen MCM, de Vries P, UytdeHaag FGCM, Osterhaus AGCM (1991) A role for $CD8^+$ class I MHC-restricted CTLs in recovery from measles: implications for the development of inactivated measles vaccine. Vaccine 91. Cold Spring Harbor Laboratory Press, Cold Spring Harbor

Van Binnendijk RS, van Baalen CA, Poelen MCM, de Vries P, Boes J, Cerundolo V, Osterhaus ADME, UytdeHaag FGCM (1992a) Measles virus transmembrane fusion protein synthesized de novo or presented in ISCOM is endogenously processed for HLA class I- and II-restricted cytotoxic T cell recognition. J Exp Med 176: 119–128

Van Binnendijk RS, Rumke HC, Van Eijdhoven MJA, Bosman A, Hirsch R, Van Loon AM, Benne CA, Van Dijk WC, Rima BK, Van der Heijden RWJ, UytdeHaag FGCM, Osterhaus ADME (1992b) A measles outbreak in vaccinated schoolaged children in the Netherlands: identification of clinically and subclinically infected children by evaluation of virus-specific antibody and T cell responses. Thesis, University of Utrecht (ISBN 90-393-0244-87)

Van Binnendijk RS, Versteeg-van Oosten JPM, Poelen MCM, Brugghe HF, Hoogerhout P, Osterhaus ADME, UytdeHaag FGCM (1993) Human HLA class I- and class II-restricted cloned cytotoxic T lymphocytes identify a cluster of epitopes on the measles virus fusion protein. J Virol 67: 2276–2284

Van Bleek GM, Natheson SG (1990) Isolation of an endogenous processed immunodominant viral peptide from the H-2Kb molecule. Nature 348: 213–216

Van Bleek GM, Nathenson SG (1991) The structure of the antigen binding groove of major histocompatibility complex class I molecules determines specific selection of self peptides. Proc Natl Acad Sci USA 88: 11032–11036

Whittle HC, Mee J, Werblinska J, Yakuba J, Onuora A, Gomwalk N (1980) Immunity to measles in malnourished children. Clin Exp Immunol 42: 144–151

Wright LL, and Levy NL (1979) Generation on infected fibroblasts of human T and non-T lymphocytes with specific cytotoxicity, influenced by histocompatibility, against measles virus-infected cells. J Immunol 122: 2379–2386

Yang Y, Water JB, Früh K, Petersen PA (1992) Proteasomes are regulated by interferon τ: implications for antigen processing Proc Natl Acad Sci USA 89: 4928–4932

Yap KL, Ada GL, McKenzie ISC (1978) Transfer of specific cytotoxic T cells protects mice inoculated with influenza viruses. Nature 273: 238–239

Yewdell JW, Bennink JR (1992) Cell biology of antigen processing and presentation to major histocompatibility complex class I molecule-restricted T lymphocytes. Adv Immunol 62: 1–123

Zhang W, Young ACM, Imarai M, Nathenson SG, Sacchettini JC (1992) Crystal structure of the major histocompatibility complex class 1 H-2 K^b molecule containing a single viral peptide: implications for peptide binding and T-cell receptor recognition. Proc Natl Acad Sci USA 89: 8403–8407

The Class I-Restricted Cytotoxic T Lymphocyte Response to Predetermined Epitopes in the Hepatitis B and C Viruses

A. CERNY[1], C. FERRARI[1,2], and F. V. CHISARI[1]

1 Introduction

Hepatitis B and C viruses (HBV and HCV) both originally identified as causative agents of transfusion-associated hepatitis share a propensity to induce acute and chronic hepatitis and hepatocellular carcinoma (TIOLLAIS et al. 1981; CHOO et al. 1989). Despite these biological similarities, the two viruses differ considerably. HBV has a partially double-stranded DNA genome of about 3200 nucleotides, infectious particles are 42 nm in size and are enveloped, and the virus belongs to the hepadnavirus family; HCV has a positive-stranded RNA genome of approximately 10 000 nucleotides, is less than 80 nm in size, is probably enveloped due to its sensitivity to organic solvents, and it is related to the Togaviridae or Flaviviridae (HOUGHTON et al. 1991).

The mechanisms whereby HBV and HCV cause acute hepatocellular injury and initiate the sequence of events leading to chronic liver disease and ultimately to hepatocellular carcinoma are not well understood. It is possible that both direct, virus-related or indirect, i.e., immunologically mediated, mechanisms may play an important role. Analysis of the direct cytopathic

[1] The Scripps Research Institute, Department of Molecular and Experimental Medicine, 10666 North Torrey Pines Road, La Jolla, CA 92037, USA
[2] Cattedra Malattie Infettive, Universita degli Studi di Parma, Parma, Italy

Current Topics in Microbiology and Immunology, Vol. 189

effects of HBV and HCV for host liver cells has been hampered due to the lack of suitable animal models and tissue culture systems. Several clinical observations underline the contribution of the host immune response to liver cell injury: (a) infection acquired early in life occurring in an immunologically immature host leads to a chronic asymptomatic carrier state; (b) chronic carriers without evidence of liver cell injury are frequent for both viruses; and (c) immunosuppression has a beneficial effect on liver cell injury in chronic hepatitis B or C (HOLLINGER 1990; ALTER 1991).

At its most fundamental level, the cellular immune response involves multimolecular interaction between antigenic peptides, HLA molecules, and T cell receptors (TCR). Unlike antigen recognition by B cell immunoglobulin receptors, the two general classes of T cells do not recognize native antigen in solution; rather they recognize short antigenic peptides that have reached the cell surface via two quite different pathways (reviewed in ROTHBARD and GEFTER 1991).

Human $CD4^+$ T cells recognize short antigenic peptides derived by proteolytic cleavage of exogenous antigen present in the antigen-binding groove of HLA class II molecules at the surface of phagocytic antigen-presenting cells (APC) such a macrophages and B cells. Effector functions of this subset of T lymphocytes include secretion of lymphokines promoting growth and differentiation of antigen specific B cells and cytotoxic T lymphocytes (CTL) as well as activation of inflammatory cells (MOSSMANN et al. 1986). $CD4^+$ T cell-mediated cytotoxicity has been described in other human viral infections such as herpesvirus (YASUKAWA and ZARLING 1984; HAYWARD et al. 1986) and will not be further discussed in this review.

Human $CD8^+$ T cells recognize short antigenic peptides (usually nine to 11 residues) in the antigen-binding groove of HLA class I molecules which are present at the surface of the cells in which their precursor (e.g., virus-derived) proteins were originally synthesized (MONACO 1992). The antigenic peptides are derived by proteolytic cleavage of endogenously synthesized antigen in the cytoplasm. The processed peptides are then bound by a family of transporter proteins (encoded within the HLA locus), which shuttle them into the lumen of the endoplasmic reticulum, where they are scanned for the presence of HLA allele-specific binding motifs by the antigen-binding domain of resident HLA class I proteins. Peptides containing the appropriate motif are bound by the corresponding HLA class I molecule, which then associates with β_2-microglobulin and moves to the cell surface as an integral membrane protein, where it can present the antigenic peptide to the appropriately rearranged TCR on a $CD8^+$ T cell. The T cell subset specificity of this interaction derives from the fact that multimolecular HLA–peptide–TCR complex is stabilized by accessory interactions such as those between the CD8 molecule on the T cell and the HLA class I molecule involved in the complex.

Due to a precedent in other systems, it has been assumed that the HLA class I-restricted, $CD8^+$ CTL response to endogenously synthesized HBV and

HCV antigens is the effector limb in pathogenesis. This was an untestable hypothesis until recently, due to the absence of the necessary reagents and experimental systems—specifically, since both viruses do not infect continuous human cell lines in tissue culture, and the only animal models of HBV (woodchuck, ground squirrel, duck) and HCV (chimpanzee) infection that could be used for such studies involve species for which the immune system is not entirely defined.

2 CTL Response to Hepatitis B Virus

2.1 Analysis of CTL in the Peripheral Blood

To circumvent these problems, we set out several years ago to develop alternative modalities to present class I-bound, processed viral antigen to the CTL repertoire in HBV-infected patients and transgenic mice. In the course of these studies, we drew upon the experience of McMichael and Townsend and colleagues (GOTCH et al. 1987), who showed that murine class I-restricted CTL responses could be induced by stimulation with synthetic peptides derived from the corresponding antigen. Accordingly, in our analysis of HBV-infected patients, we stimulated peripheral blood mononuclear cells (PBMC) with pools of randomly designed, 15- to 20-residue synthetic HBV-derived peptides followed by restimulation with autologous Epstein-Barr virus (EBV)-transformed B cell lines that had been stably transfected with vectors that express the corresponding HBV-encoded protein. Although this process was relatively inefficient, it yielded HBV-specific CTL lines and clones that were assessed for phenotype, HLA restriction, recognition of endogenously synthesized antigen, and antigenic fine specificity. In this manner, we defined two HLA-A2-restricted CTL epitopes in the viral nucleocapsid ($HBcAg_{18-27}$) and envelope ($HBsAg_{335-343}$) proteins (GUILHOT et al. 1992; BERTOLETTI et al. 1991; PENNA et al. 1991; BERTOLETTI et al. 1993; NAYERSINA et al. 1993) and two additional completely overlapping epitopes within $HBcAg_{141-151}$ that, remarkably, are restricted by two independent HLA class I molecules (HLA-A31 and HLA-Aw68) in patients with acute viral hepatitis (MISSALE et al. 1993). In contrast to these results in patients with acute hepatitis, it is extremely interesting and potentially very important that thus far we have not detected a CTL response to any of these epitopes in the peripheral blood of a large number of HLA class I-matched patients with chronic hepatitis (BERTOLETTI et al. 1993; MISSALE et al. 1993).

It is interesting that the CTL epitope located between HBcAg residues 141 and 151 (STLPETTVVRR; MISSALE et al. 1993) completely overlaps a critical domain in the viral nucleocapsid protein that is essential for its nuclear localization and genome packaging functions (ECKHARDT et al. 1991; NASSAL 1992), as well as processing of the precore protein (SCHLICHT and

SCHALLER 1989). Because of this feature, the CTL response to this epitope could be especially effective at viral clearance since viral mutations in this region might be lethal for the virus itself. This study yielded an unexpected bonus for investigators interested in the molecular basis of HLA class I–antigenic peptide interaction, because the CTL response to this epitope was found to be dually restricted by the HLA-A31 and HLA-Aw68 alleles and, even more unexpectedly, both responses are focused on precisely the same 11-residue sequence. Perhaps most importantly, we discovered, by alanine substitution and competition analysis, that both of these class I alleles utilize the same HLA-binding motif, which consists of a threonine in position 2, a leucine in position 3, and an arginine at the carboxy terminus of an 11-residue peptide. Furthermore, one patient yielded two independent CTL clones whose antigenic fine specificity differed according to the HLA allele that restricted the response (i.e., glutamic acid 145 and valine 149 are epitopic residues for the HLA-A31-restricted clone, and the two threonine residues at positions 146 and 147 are seen by the TCR of the clone restricted by HLA-Aw68), once again emphasizing the multispecificity and polyclonality of the CTL response to HBV during acute hepatitis. Finally, the influence of peptide–class I interactions on TCR recognition was demonstrated by the observation that neither clone could recognize its cognate antigenic peptide when presented by the alternate restriction element. In addition to the substantial fundamental implications of this discovery, the identification of broadly presented epitopes such as this could have significant practical value for the development of a peptide-based CTL vaccine for termination of chronic HBV infection.

As we pursued these studies using randomly designed peptides, we discovered that the two HLA-A2-restricted CTL epitopes that were identified in this way were short peptides (9-mer and 10-mer) that displayed the ideal HLA-A2.1 allele-specific binding motif (xLxxxxxxV) that Rammensee and his collaborators (FALK et al. 1991) had independently found by eluting naturally synthesized cellular peptides from the HLA-A2.1-binding groove. Accordingly, a large panel of peptides was designed corresponding to ideal and alternative HLA-A2.1-binding motifs present within the HBV envelope, nucleocapsid, polymerase, and X proteins. We subsequently screened these peptides by stimulation of PBMC from HLA-A2-positive patients with acute hepatitis and identified additional HLA-A2-restricted CTL epitopes. During the course of these studies we learned that CTL lines that recognize endogenously synthesized antigen could be established simply by repeated stimulation with peptide rather than with stable, autologous, HBV transfectants, thereby greatly simplifying the CTL activation procedure.

Some of these peptides have already been used to examine the HBV-specific CTL response in several HLA-A2-positive patients with acute hepatitis. These studies demonstrated that some epitopes are recognized by most of the acutely infected patients, while the others are recognized by only a few, suggesting that factors other than the HLA-binding motif influence immunogenicity. It is also clear that some acutely infected patients respond to many of the epitopes, while the response of others is much more restricted. However, we will need to study an expanded panel of patients with an extended panel of epitopes before a definitive epitope recognition hierarchy is established.

Another very interesting and important observation is that some acutely infected patients studied with this panel of peptides failed to respond to any of them, similar to chronically infected patients and uninfected normal controls. It is possible, therefore, that these actuely infected nonresponders might ultimately fail to clear the virus. Alternatively, it is possible that they actually have chronic hepatitis which presented clinically with a flare that was misdiagnosed as acute hepatitis; or they may produce a perfectly good response to other epitopes, restricted by HLA-A2 or any of their other class I alleles, that we have not yet tested. A less likely possibility is that they are infected by variant viral genomes that do not encode any of these epitopes. To examine these clinically important alternatives, it will be necessary to follow these patients sequentially, to extend the analysis to new epitopes, and to determine the amino acid sequence in the region of these epitopes that is encoded by the viral genomes with which each of these patients is infected. It is exciting to think that we might possibly have identified a marker to distinguish patients who are destined to clear the virus from those who will become chronically infected by HBV. If this proves to be true in our future studies, not only will we have gained an important insight into the biology of viral persistence, but we may have developed a means to select patients who might benefit from early intervention with antiviral or immunomodulatory therapy designed to prevent the establishment of the chronic HBV carrier state.

Perhaps the most important fact to emerge from these studies is that the HBV-specific CTL response is remarkably polyclonal and multispecific within individual patients who succeed in clearing the virus, presumably conferring a high degree of protection against this dangerous viral pathogen. The diversity of the response also minimizes the likelihood that a mutation at any one of these epitopes would confer a selective survival advantage to the mutant virus. However, the possibility that such mutants may be actually selected in a minor proportion of patients who express a weaker and less heterogenous CTL response, such as patients with chronic hepatitis B, cannot be ruled out at present. Escape mutants in this setting would not be the original cause of virus persistence, but only a consequence of a prior defect in the antiviral immune response. The basis for the difference in the immunogenicity of these epitopes and the differential responsiveness of the patients is not clear at present, nor is the clinical significance of a strong response to many epitopes in certain patients versus a weak response to one or a few epitopes in others.

The fact that this sequential stimulation strategy worked so well, whereas many prior attempts failed to detect HBV-specific CTL in freshly isolated peripheral blood lymphocytes PBL without prior stimulation or following stimulation with stably transfected autologous B cell lines (BCL) without preliminary expansion with specific peptides, suggests that the HBV-specific CTL precursors are probably present in the peripheral blood compartment at very low frequency. Despite the effectiveness of this strategy in patients with acute hepatitis, however, we have rarely seen a response to these epitopes in patients with chronic hepatitis using the same stimulation conditions. The reasons for this are unclear. Possible mechanisms include: (a) the CTL response may be absolutely dependent on T cell help derived from a concomitant HBV-specific, $CD4^+$ T cell response, which we have shown is universally present in patients who clear the virus and absent in those who do not (FERRARI et al. 1991); (b) there may be a "hole" in the T cell repertoire for HBV-encoded antigens in patients with chronic hepatitis, due to the induction of immunological tolerance; (c) these patients may have a defect in processing of HBV antigens such that they do not cleave or transport the approriate peptide subunits to the appropriate HLA class I molecule; (d) they may express variant HLA class I alleles that cannot bind the peptides; (e) they may be infected by viral variants that do not contain the epitopes needed for viral clearance; and (f) they may generate a suppressive T cell response to HBV antigenic stimulation. Further studies will be needed to discriminate among these possibilities.

These data also suggest that HBV mutants that result in deletion of a single epitope or even an entire protein should not lead to escape of this antigenically complex virus from immune recognition unless the CTL response in that patient is functionally defective, and thus monoclonal or oligoclonal, or unless the mutation also confers a growth advantage on the mutant virus. Since we have shown that the HBV major, middle, and large envelope proteins are all good targets of the HBsAg-specific CTL response (NAYERSINA et al. 1993), just as the two nucleocapsid proteins are good targets of the HBcAg-specific CTL response (BERTOLETTI et al. 1991; MISSALE et al. 1993), even mutant viruses that lose the ability to synthesize one or more of these proteins in their entirety should not escape CTL recognition in the setting of a normal immune response, because the epitopes would still be generated by processing of the remaining protein(s). If, for some reason, however, a mutant can escape recognition in the original host, it should be irrelevant to subsequent hosts unless they have exactly the same HLA haplotype as the proband, a very rare situation indeed.

The principal strength of this peptide stimulation strategy is that it permits analysis of the CTL response to pathogens whose genome has been cloned and sequenced, but which are not infectious for autologous or surrogate stimulator/target cells in vitro. Indeed, as will be discussed below, we have recently extended the strategy for analysis of the HLA-A2-restricted CTL response to HCV with very encouraging results, and we will explore the CTL response restricted by other class I alleles as their binding motifs become defined.

There are, however, numerous limitations to this strategy which deserve attention. First, it requires knowledge of the amino acid sequence of the viral protein(s) in question; this is not needed for the initial identification of the CTL response to viral pathogens that are infectious in vitro. Second, to be most efficient, the strategy must be limited to epitopes presented by common HLA alleles whose binding motifs have been defined. Third, it has the potential to miss CTL responses to nonconserved epitopes, since the patient's CTL may be primed in vivo to an epitope that differs at one or more positions from the peptide used for in vitro expansion and screening. This problem can be readily overcome by focusing on highly conserved peptides carrying the HLA-binding motif of interest; but, correspondingly, this requires that the amino acid sequences of multiple independently cloned viral isolates must be known. Fourth, for these reasons plus the fact that not all peptides containing a given HLA-binding motif actually bind the restriction element with equal affinity, the process is inefficient and expensive, requiring the synthesis of many peptides, only a small proportion of which will prove to be CTL epitopes. This can be overcome to some extent by prescreening the peptides for HLA-binding affinity and focusing on the conserved peptides with the highest affinity for the corresponding allele. Obviously, this interposes an intermediate step in the strategy. It is worthwhile to do so, however, since in preliminary studies done in collaboration with R. Kubo and A. Sette (Cytel Corp., La Jolla, CA) we have found that only approximately 25% of HBV-derived peptides with the HLA-A2-binding motif actually bind to this molecule with reasonably high affinity. Importantly, slightly more than one third of these high-affinity peptides have proven to be epitopes recognized by CTL in acutely infected patients, while virtually none of the peptides with low binding affinity are reactive in this system. Finally, it is possible that this peptide stimulation strategy could lead to the in vitro induction of a primary CTL response that does not reflect in vivo priming by the pathogen. Indeed, we have recently shown in healthy uninfected donors that this can occur. While this represents a potential, but surmountable problem for the assessment of the pathogenetic significance of a given CTL response in infected individuals, it represents a major opportunity if one is interested in defining the CTL repertoire to any intracellular pathogen.

In vivo- and in vitro-primed CTL can be distinguished firstly on the basis of their precursor frequency in the peripheral blood and secondly according to the surface expression of CD45 isoforms (ALEXANDER et al. 1992). We found a ten to 100-fold higher precursor frequency of HLA-A2-restricted CTL precursor specific for a single HBV epitope in the peripheral blood of patients with chronic hepatitis B than in normal subjects. Depletion of CD45 RO^+ PBMC reduces the CTL precursor frequency of chronically infected patients to the level of normal subjects, suggesting that in vivo-primed CTL precursors are CD45 RO^+ (A. Cerny, M. A. Brothers, and F. V. Chisari, unpublished observation).

Obviously, all of the foregoing comments carry important implications for understanding the role of the CTL response in viral pathogenesis and immunobiology. Additionally, these observations are also relevant for the possible development of peptide-based CTL vaccines. In this context, A. Vitiello and R. Chesnut (Cytel) have recently shown in preliminary studies, that a formulation containing the highly conserved HLA-A2-restricted CTL epitope we identified in the HBV nucleocapsid protein ($HBcAg_{18-27}$) is immunogenic in HLA-A2.1 transgenic mice (A. Vitiello, and R. Chesnut, personal communication). Pilot studies to evaluate the immunogenicity of this formulation in HLA-A2-positive humans and chronic HBV carriers are currently underway.

2.2 Analysis of Intrahepatic CTL

Immunohistochemical studies on the intrahepatic inflammatory infiltrates in chronic hepatitis associated with HBV infection demonstrate an increase in $CD8^+$ mononuclear cells predominantly in the infiltrates of periportal areas (piecemeal necrosis; YANG et al. 1988). This observation was corroborated by studies using different cloning strategies to derive T cell clones from liver biopsies that showed that most of the clones were $CD8^+$ and cytotoxic (HOFFMANN et al. 1986; MEUER et al. 1988). The information derived from these studies remains limited, due to the unavailability of tools to further study antigen specificity at the time when they were performed.

It is very interesting that two groups of investigators have demonstrated the presence of $CD8^+$ CTL in the intrahepatic inflammatory infiltrates in patients with chronic HBV infection using recombinant HBcAg and HBsAg particles for in vitro stimulation. Ferrari and colleagues have demonstrated that hepatitis core antigen is an important target of the intrahepatic TCR (FERRARI et al. 1987) and that $CD4^+$ and $CD8^+$ HBcAg-specific T cells are present in the intrahepatic inflammatory cell infiltrate during chronic HBV infection. While the $CD8^+$ T cells were shown to be cytotoxic in antigen-nonspecific cytotoxicity assays, the systems needed to demonstrate antigen-specific killing were not available when these studies were done.

In contrast, Barnada and colleagues have demonstrated that an epitope within the pre-S2 domain of the viral envelope protein was also recognized by intrahepatic $CD4^+$ and $CD8^+$ T cell clones during chronic HBV infection. With the use of synthetic peptides, an HLA-A3-restricted epitope could be precisely mapped to pre-S2 residues 120–134 (BARNABA et al. 1989). Interestingly, exogenously added HBV envelope antigen was also able to sensitize antigen-specific B cells for CTL-mediated lysis (BARNABA et al. 1990). Both of these observations suggest that class I-restricted, $CD8^+$ CTL reponses can be induced by certain exogenous particulate antigens, presumably by entry into the class I processing pathway of professional APC. While it cannot be proven that this occurred in the aforementioned study of infected patients, it must have occurred in the study by Berkower's group in which class I-restricted T cells specific for pre-S1 residues 21–28 were identified in the peripheral blood of HBsAg subunit vaccine recipients (JIN et al. 1988). It is remarkable that in addition to the classical pathway for induction of class I-restricted CTL by endogenously synthesized antigen, this generally unappreciated alternative pathway appears to be utilized for the CTL response to multiple HBV-derived antigens, perhaps underscoring the importance of elimination of this dangerous viral pathogen by the host.

3 CTL Response to Hepatitis C Virus

3.1 Analysis of CTL in the Peripheral Blood

In an early study performed before the identification of HCV as the major cause of non-A, non-B hepatitis in 1989, IMAWARI et al. (1989) described a human $CD8^+$ T cell clone derived from the peripheral blood of a patient with non-A, non-B hepatitis cytotoxic for autologous and allogeneic hepatocytes from patients with non-A, non-B hepatitis. Due to the lack of information on the agent responsible for non-A, non-B hepatitis, the fine specificity of this CTL clone was undefined.

In an attempt to define the molecular targets of the HCV-specific CTL response in the peripheral blood of chronically infected patients, we adopted an in vitro expansion strategy similar to the one we had previously used for HBV using HCV-derived synthetic peptides (Cerny et al., manuscript in preparation). We focused our analysis on CTL epitopes restricted by HLA-A2 for which the binding motif xLxxxxxxV or xLxxxxxxxV had been previously identified (FALK et al. 1991). The HCV-1 amino acid sequence was then scanned for the presence of the HLA-A2.1-binding motif and 53 peptides were identified for synthesis. Two peptides were derived from the core region, two from E1, six from E2/NS1, nine from NS2, nine from NS3, ten from NS4, and 15 from NS5.

Eight subjects positive for the HLA-A2 allele with hepatitis C infection were identified, PBMC were stimulated individually with the entire panel of 53 peptides, and cultures were tested after initial expansion for peptide-specific CTL activity. Six patients (C-1 to C-6) had biopsy-proven chronic active hepatitis associated with HCV infection, and subjects H-1 and H-2 had no signs of liver disease. Table 1 summarizes our results. Peptide-specific cytotoxic T cell activity was observed against seven out of 53 peptides. Four of the eight subjects showed CTL responses to at least one of the peptides. Subject C-2 responded to five peptides, two of which are derived from HCV core and one from NS3, NS4, and NS5 respectively. Subject C-3 responded to four peptides, including HCV $core_{178-187}$, but not HCV $core_{131-140}$. C-5, in

Table 1. Summary of hepatitic C virus (HCV) peptide specific CTL responses

HCV	Amino acid residues	Patients responding[a]
Core	131–140	C-2, C-5
Core	178–187	C-2, C-3
NS3	1169–1177	C-3
NS3	1406–1415	C-2, C-3, C-5
NS4	1789–1797	C-2, H-1
NS4	1807–1816	C-3
NS5	2252–2260	C-2

[a] Peripheral blood mononuclear cells were stimulated with the panel of 53 peptides and cultures were tested after initial in vitro expansion for peptide-specific CTL activity. A difference in the specific lysis of peptide-pulsed target cells and nonpulsed target cells of 15% at an effector-to-target cell ratio of 40/1-1/80 was considered to represent a positive CTL response and was confirmed by additional rounds of restimulation and subsequent cloning.

contrast, recognized HCV $core_{131-140}$, but not HCV $core_{178-187}$. Subject H-1 responded to only one peptide: $NS4_{1789-1797}$. Several of the peptides were found to be stimulatory for more than one patient, probably reflecting a higher degree of immunogenicity.

Four of the subjects (C-1, C-4, C-6 and H-2) did not show any significant induction of CTL activity with this panel of peptides. Both patients without overt liver disease as well as patients with chronic liver disease belong to this subset.

Further characterization of the CTL effector cells demonstrated that they were $CD8^+$, HLA-A2 restricted, and, in the case of $NS3_{1406-1415}$-specific CTL, capable of recognizing endogenously synthesized antigen (data not shown).

Due to the limited number of observations and the cross-sectional design of this study, it is at this time difficult to correlate the CTL response found with the clinical status or the virological parameters of the subjects studied. It is noteworthy that three patients with biopsy-proven chronic active hepatitis (CAH) did not have detectable CTL in their peripheral blood. Several possible explanations may account for this. First, the CTL precursor frequency specific for the peptides used in peripheral blood may be below our detection limit in these patients. Second, it may be that the CTL precursor frequency fluctuates in a given subject and the cross-sectional format of our pilot study fails to reveal this. Third, the patient may have a variant HLA-A2 allele cross-reacting serologically, but having a different HLA-binding motif. Several variants of HLA-A2 have been described that cross-react serologically, but not at the level of CTL restriction (Lopez and Castro 1989; Rötzschke et al. 1992). Fourth, HCV-specific CTL may be sequestered at the site of infection, i.e., the liver, and not detectable in the peripheral blood. Finally, the patient may be infected with a variant HCV isolate containing amino acid substitutions

Table 2. Representation of the hepatitis C virus (HCV) CTL epitope sequences in different HCV subtypes

HCV	Amino acid residues	HCV Subtype				
		I[a]	II	III	IV	ND
Core	131–140	3[b]/3[c]	8/8	1/3	2/2	7/8
Core	178–187	3/3	1/11	0/3	2/2	2/8
NS3	1169–1177	2/3	0/5	0/1	0/1	0/1
NS3	1406–1415	4/5	0/5	0/1	0/1	0/1
NS4	1789–1797	3/3	0/5	0/1	0/1	0/1
NS4	1807–1816	3/3	5/5	0/1	0/1	1/1
NS5	2252–2260	3/3	0/5	0/1	0/1	0/1

ND, not determined.
[a] Subtype of HCV as described (OKAMOTO et al. 1992).
[b] Number of sequences that show no amino acid substitutions within a given epitope.
[c] Total number of sequences deposited in GenEMBL covering a given epitope.

within the CTL epitopes tested that we derived from the HCV-1 sequence. In fact, HCV, a positive-stranded RNA virus, displays considerable sequence variability among reported isolates. The representation of the peptide sequences within the HCV subtypes whose sequence is available through the Gen EMBL data base is shown in Table 2.

The peptide sequence $NS3_{1406-1415}$, which was recognized by CTL from three subjects, is present in four out of five HCV-I subtypes predominant in the US and Europe. The fifth isolate, HCV-H, differs only with respect to one conservative Ileu to Val substitution at position 7. This makes it an interesting epitope for further study and a potential candidate for a peptide-based HCV vaccine as discussed above for HBV.

It is noteworthy that the peptide $NS3_{1169-1177}$ is part of the viral serine proteinase close to residue Ser 1165, which is one of the three catalytic site residues (GRAKOUI et al. 1993). Changes within that sequence may thus affect proteinase activity and compromise viral replication. This sequence is a biologically particularly important molecular target for the host immune system.

The link between HCV infection and the presence of autoantibodies is well established. One set of autoantibodies was originally defined by its reactivity with cryostat sections of rat liver and kidney and called LKM antibodies. A subset of LKM antibodies termed LKM-1 recognize epitopes within the human cytochrome p450IID6 protein and are associated with autoimmune hepatitis type II (MANNS et al. 1991). A significant proportion of patients with the disorder, particularly in southern Europe, are infected with HCV (LENZI et al. 1991). The presence of other microsomal autoantigens has been suggested in these patients (LENZI et al. 1991).

We therefore undertook an amino acid sequence comparison using the SwissProt data base and identified two human cytochrome p450 sequences related to HCV $core_{178-187}$, i.e., p450IIA6 and p450IIA7 (NEBERT and

Table 3. Hepatitis C virus (HCV) core 178–187 and related human cytochrome p450 amino acid and HCV sequences

Origin	Residues	Sequence[b]
HCV-1 core	178–187	L L A L L S C L T V
Cytochrome p450 IIA6	9–18	• V • • • V • • • •
Cytochrome p450 IIA7	9–18	• V • • • A • • • •
HCV subtype II ($n = 8^a$)	178–187	• • • • • • • • • I
HCV subtype II ($n = 2^a$)	178–187	• • • • • • • • • T
HCV J5 (III)	178–187	• • • • • • • I • •
HCV J6 (III)	178–187	• • • • • • • I • T
HCV KF	178–187	• S • • M • • • • T
HCV 476	178–187	• S • • M • • • • A

[a] Designates the number of isolates containing the amino acid sequence indicated.
[b] Amino acid sequence comparison was performed using the SwissProt and GenEMBL data bases and identified two human cytochrome p450 sequences related to HCV $core_{178-187}$. Identical amino acids are shown as dots. The amino acid single letter code is used.

NELSON 1991; Table 3). They display an eight-amino acid sequence identity, one conservative substitution (Leu to Val), and one nonconservative substitution (Ser to Val and Ser to Ala, respectively). Both cytochrome p450 mono-oxygenase sequences contain the HLA-A2 anchor residues, since both Leu and Val can serve as anchor residues at position 2 (R. Kubo and A. Sette personal communication). HCV sequences deposited in Gen EMBL show a sequence variability in the HCV $core_{178-187}$ sequence as well (Table 3). Given the fact that both HCV and cytochrome p450 sequences are polymorphic, coincidental sequence identity may occur in an HCV-infected individual. This suggests the possibility of molecular cross-reactivity at the level of CTL between HCV and autoantigens expressed on hepatocyte class I molecules, a hypothesis that we are currently addressing experimentally. Such a scenario could also account for geographic differences in the incidence of HCV-associated autoimmunity, assuming sequence identity within a CTL epitope between a prevalent HCV and cytochrome p450 genotype in a give population. Sequence identity was, however, not required for cross-recognition of a D^b-restricted mouse CTL epitope in HCV NS5 (aa 2422–2437). Three peptides derived from variant HCV isolates containing different single amino acid substitutions were also able to sensitize target cells for CTL-mediated lysis (SHIRAI et al. 1992).

The finding that multiple HCV-derived peptides serve as CTL epitopes restricted by a single class I restriction element underlines the multispecificity of the human cellular immune response against HCV. This observation parallels results obtained in a study on the $CD4^+$ proliferative TCR of PBMC to recombinant HCV-derived proteins: all viral proteins were found to be immunogenic in that study, and responsiveness was detectable in more than half of seropositive individuals (BOTARELLI et al. 1993).

The general experimental strategy reviewed here, i.e., selection of candidate CTL epitopes using a HLA-binding motif applied to a known sequence and subsequent screening of peptides for CTL induction, is expected to be a useful approach for the analysis of the CTL response in other systems as well.

3.2 Analysis of Intrahepatic CTL

The analysis of the intrahepatic CTL response has the advantage of reflecting the local immune response at the site of infection more faithfully. The requirement to perform a liver biopsy and the limited amount of tissue obtained with the associated sampling error are inherent problems associated with this approach. The local response may also be diluted due to the local recruitment of antigen-nonspecific inflammatory cells. On the other hand, HCV-spcific CTL may be sequestered at the site of infection, i.e., the liver, and not detectable in the peripheral blood.

Using an in vitro expansion protocol involving the use of a bispecific anti-CD3/anti-CD4 antibody allowing for the selective expansion of $CD8^+$ cells, Koziel and coworkers described class I-restricted $CD8^+$ CTL clones derived from liver biopsies from the two subjects evaluated recognizing epitopes within E1 and E2(NS1)/NS2 (Koziel et al. 1992). One epitope in E1 (aa 235–242) is restricted by HLA-B35 and a second epitope in NS2 (aa 826–837) is restricted by HLA-A29; a third epitope situated within E2(NS1)/NS2 is restricted by HLA-A2. Attempts to detect CTL in the peripheral blood of these same patients using a similar experimental approach were unsuccessful, suggesting a compartmentalization of the antiviral CTL response at the site of infection and tissue damage (B. Walker and M. Koziel, personal communication).

Another study used experimental HCV infection of chimpanzees, which allows for the study of all phases of the natural history of HCV infection. NS3 was found to be the dominant target of intrahepatic CTL-mediated immunity. The two animals studied had NS3-specific CTL restricted by distinct class I restriction elements that in one animal could be detected as long as 28 weeks postinfection (C. M. Walker, personal communication). Important information regarding the protective versus pathogenic potential of CTL can be expected from such studies.

4 Comparison of the CTL Response to Hepatitis B and C Viruses

It is interesting to note that in contrast to chronic HCV infection, this and other laboratories have not been successful in consistently identifying cytotoxic T cell responses in the peripheral blood of patients with chronic

HBV infection (MISSALE et al. 1993). This discrepancy may be due to a more vigorous antiviral CTL response in HCV infection as compared to HBV.

Alternatively, the CTL response to HBV may be more compartmentalized and restricted to the liver. Both HBV and HCV have been found to be detectable in PBMC of chronically infected patients (BOUFFARD et al. 1992; BAGINSKI et al. 1991). Subtle differences in tropism and expression, however, such as an enhanced presence of replicating virus in circulating professional APC (dendritic cells, monocytes, and B cells) could according to this hypothesis be the basis of the more systemic nature of the immune response against HCV.

5 Conclusions

The past several years have seen the definition of the molecular events underlying HLA class I restriction and have allowed a molecular analysis of CTL specificity. This technology applied to HBV and HCV pathobiology has and will yield a wealth of detailed information useful in many areas of heptitis research such as to elucidate pathogenesis, to refine diagnostics, and to develop new therapeutic concepts.

Autoimmunity due to molecular mimicry as suspected by clinical observations and by the finding of highly related amino acid sequences, illustrated by the case of the HCV CTL epitope $core_{178-187}$ and cytochrome p450, provide testable experimental hypotheses that may lead to a molecular definition of CTL-mediated autoimmunity.

The cytotoxic T cell response to both HBV and HCV is surprisingly multispecific, involving epitopes in different structural and nonstructural proteins and also polyclonal, especially considering the fact that the studies done with HLA-A2-binding peptides only reveal a small segment of the class I-restricted CTL response. Similar observations have been made in other human viral diseases such as human immunoreactivity virus (HIV; WALKER et al. 1989), varicella zoster virus (VZV; ARVIN 1992), herpes simplex virus (HSV; TIGGES et al. 1992), and cytomegalovirus (CMV; RIDDELL et al. 1991). Viral escape due to mutation of a CTL epitope as reported in a monoclonal mouse system is not likely to be relevant for viral persistence in the above human viral diseases (PIRCHER et al. 1990), except in individuals whose entire CTL response is focused on a single viral determinant, if such patients can be identified. If this occurs, any mutant virus should be irrelevant for the next host in view of the diversity of human HLA alleles, unless that mutation also confers a positive growth advantage to the virus that is independent of any immune selection pressure. This, indeed, might be the basis for the relative prevalence of an HLA-All-restricted EBNA-4 epitope loss mutant of EBV recently reported to occur in highly HLA-A11-positive populations (De

CAMPOS-LIMA et al. 1993). The issue remains controversial, however, since indirect evidence for a role of viral escape mutants has been reported in a longitudinal study of HIV-infected hemophiliac donors documenting the emergence of mutants with changes in CTL epitopes within HIV gag abolishing recognition by CTL from the same individual (Phillips et al. 1991). However, no selection of mutations in an immunodominant CTL epitope in simian immunodeficiency virus (SIV) gag could be detected when compared to the flanking non-epitope-coding region in a study done in rhesus monkeys (CHEN et al. 1992).

The key question remains unanswered, however: how can a virus persist under a polyclonal, multispecific immune attack? Apart from escape by mutation of epitopes, many different examples of modulation of the immune response in various viral infections have been reported and reviewed recently (GOODING 1992). They include interference with class I antigen translocation, cytokine production and function, and lymphocyte signal transduction.

The experimental strategy reviewed in this paper, i.e., selection of candidate CTL epitopes using an HLA-binding motif applied to a virus sequence and subsequent screening of peptides for CTL induction, may be generally useful for the analysis of the CTL response in other systems where APC that synthesize the target protein of interest are not available.

Acknowledgements. This work was supported by grants RO1 AI 20001, AI26626, and RR 00833 from the National Institutes of Health and funds from Cytel Corporation, La Jolla, CA. Andreas Cerny was supported by a fellowship of the Schweizerische Stiftung für Medizinisch Biologische Stipendien. We thank R. Kubo, A. Sette, A. Vitiello, R. Chestnut, C. M. Walker, and B. Walker for sharing unpublished data, our clinical collaborators A. Redeker, J. Person, and J. McHutchison for referral of patients, L. Wilkes and K. Cox for the processing of blood samples, P. Fowler and M.A. Brothers for technical assistance, and B. Weier for secretarial help. This is manuscript number 8133-MEM from the Scripps Research Institute.

References

Alexander D, Shiroo M, Robinson A, Biffen M, Shivnan E (1992) The role of CD45 in T-cell activation—resolving the paradoxes? Immunol Today 13: 477–481

Alter MJ (1991) Epidemiology of community-acquired hepatitis C. In: Hollinger FB, Lemon SM, Margolis HS (ed) Viral hepatitis and liver disease. Williams and Wilkins, Baltimore, pp 410–413

Arvin AM (1992) Cell-mediated immunity to varicella-zoster virus. J Inf Dis 166 (Suppl 1): S35–41

Baginski I, Chemin I, Bouffard P, Hantz O, Trepo C (1991) Detection of polyadenylated RNA in hepatitis B virus-infected peripheral blood mononuclear cells by polymerase chain reaction. J Inf Dis 163: 996–1000

Barnaba V, Franco A, Alberti A, Balsano C, Benvenuto R, Balsano F (1989) Recognition of hepatitis B virus envelope proteins by liver-infiltrating T lymphocytes in chronic HBV infection. J Imunol 143: 2650–2655

Barnaba V, Franco A, Alberti A, Benvenuto R, Balsano F (1990) Selective killing of hepatitis B envelope antigen-specific B cells by class I-restricted, exogenous antigen-specific T lymphocytes. Nature 345: 258–260

Bertoletti A, Chisari FV, Penna A, Guilhot S, Galati L, Fowler P, Vitiello A, Chesnut RC, Fiaccadori F, Ferrari C (1993) Definition of a minimal optimal cytotoxic T cell epitope within the hepatitis B virus nucleocapsid protein. J Virol 67: 2376–2380

Bertoletti A, Ferrari C, Fiaccadori F, Penna A, Margolskee R, Schlicht JH, Fowler P, Guilhot S, Chisari FV (1991) HLA class I restricted human cytotoxic T cells recognize endogenously syn- thesized hepatitis B virus nucleocapsid antigen. Proc Natl Acad Sci USA 88: 10445–10449

Botarelli P, Brunetto MR, Minutello MA, Calvo P, Unutmaz D, Weiner AJ, Choo Q-L, Shuster JR, Kuo G, Bonino F, Houghton M, Abrignani S (1993) T-lymphocyte response to hepatitis C virus in different clinical courses of infection. Gastroenterology 104: 580–587

Bouffard P, Hayashi PH, Acevedo R, Levy N, Zeldis JB (1992) Hepatitis C virus is detected in a monocyte/macrophage subpopulation of peripheral blood mononuclear cells of infected patients J Inf Dis 166: 1276–1280

Cerny A, McHutchison JG, Pasquinelli C, Brothers MA, Fowler P, Houghton M, Chisari FV (in preparation)

Chen ZW, Shen L, Miller MD, Ghim SH, Hughes AL, Letvin NL (1992) Cytotoxic T lymphocytes do not appear to select for mutations in an immunodominant epitope of simian immunodeficiency virus gag. J Immunol 149: 4060–4066

Choo QL, Kuo G, Weiner AJ, Overby LR, Bradley DW, Houghton M (1989) Isolation of a cDNA clone derived from a blood-borne non-A, non-B viral hepatitis genome. Science 244: 359–362

De Campos-Lima P-O, Gavioli R, Zhang Q-J, Wallace LE, Dolcetti R, Rowe M, Rickinson AB, Masucci MG (1993) HLA-A11 epitope loss of Epstein-Barr virus from a highly A11 + population. Science 260: 98–100

Eckhardt SG, Milich DR, McLachlan A (1991) Hepatitis B virus core antigen has two nuclear localization sequences in the arginine-rich carboxyl terminus. J Virol 65: 575–582

Falk K, Roetzschke O, Stevanovic S, Jung G, Rammensee H-G (1991) Allele-specific motifs revealed by sequencing of self-peptides eluted from MHC molecules. Nature 351: 290–296

Ferrari C, Penna A, Giuberti T, Tong MJ, Ribera E, Fiaccadori F, Chisari FV (1987) Intrahepatic, nucleocapsid antigen specific T cells in chronic active hepatitis. Br J Immunol 139: 2050–2058

Ferrari C, Bertoletti A, Penna A, Cavalli A, Valli A, Missale G, Pilli M, Fowler P, Giuberti T, Chisari FV, Fiaccadori F (1991) Identification of immunodominant T cell epitopes of the hepatitis B virus nucleocapsid antigen. J Clin Invest 88: 214–222

Gooding LR (1992) Virus proteins that counteract host immune defenses. Cell 71: 5–7

Gotch F, Rothbard J, Howland K, Townsend A, McMichael A (1987) Cytotoxic T lymphocytes recognize a fragment of influenza matrix protein in association with HLA-A2 Nature 326: 881–882

Grakoui A, McCourt DW, Wychowski C, Feinstone SM, Rice CM (1993) Characterization of the hepatitis C virus-encoded serine proteinase: determination of proteinase-dependent polyprotein cleavage sites. J Virol 67: 2832–2843

Guilhot S, Fowler P, Portillo G, Margolskee RF, Ferrari C, Bertoletti A, Chisari FV (1992) Hepatitis B virus (HBV)-specific cytotoxic T cell response in humans: production of target cells by stable expression of HBV-encoded proteins in immortalized human B cell lines. J Virol 66: 2670–2678

Hayward AR, Pontesilli O, Herberger M, Laszlo M, Levin M (1986) Specific lysis of varicella zoster virus-infected B lymphoblasts by human T cells. J Virol 58: 179–184

Hoffmann RM, Pape GR, Rieber P, Eisenburg J, Döhrmann J, Zachoval R, Paumgartner G, Riethmüller G (1986) Cytolytic T cell clones derived from liver tissue of patients with chronic hepatitis B Eur J Immunol 16: 1635–1638

Hollinger FB (1990) Hepatitis B virus. In: Fields BN, Knipe DM et al. (eds) Virology. Raven, New York, pp 2171–2236

Houghton M, Richman K, Berger K, Lee C, Dong C, Overby L, Weiner A, Bradley D, Kuo G, Choo Q-L (1991) Hepatitis C virus (HCV), a relative of the pestiviruses and flaviviruses. In: Hollinger FB, Lemon SM, Margolis HS (eds) Viral hepatitis and liver disease. Williams and Wilkins, Baltimore, pp 328–333

Imawari M, Nomura M, Kaieda T, Moriyama T, Oshimi K, Nakamura I, Gunji T, Ohnishi S, Ishikawa T, Nakagama H, Takaku F (1989) Establishment of a human T cell clone cytotoxic for both autologous and allogeneic hepatocytes from chronic hepatitis patients with type non-A, non-B virus. Proc Natl Acad Sci USA 86: 2883–2887

Jin Y, Shih W-K, Berkower I (1988) Human T cell response to the surface antigen of hepatitis B virus (HBsAg). J Exp Med 168: 293–306
Koziel MJ, Dudley D, Wong JT, Dienstag J, Houghton M, Ralston R, Walker BD (1992) Intrahepatic cytotoxic T lymphocytes specific for hepatitis C virus in persons with chronic hepatitis. J Immunol 149: 3339–3344
Lenzi M, Johnson PJ, McFarlane IG, Ballardini G, Smith HM, McFarlane BM, Bridger C, Vergani D, Bianchi FB, Williams R (1991) Antibodies to hepatitis C virus in autoimmune liver disease: evidence for geographical heterogeneity. Lancet 338: 277–280
Lopez de Castro JA (1989) HLA-B27 and HLA-A2 subtypes: structure, evolution and function. Immunol Today 10: 239–246
Manns MP, Griffin KJ, Sullivan KF, Johnson EF (1991) LKM-1 autoantibodies recognize a short linear sequence in p450IID6, a cytochrome p-450 monooxygenase J Clin Invest 88: 1370–1378
Meuer SC, Moebius U, Manns MM, Dienes HP, Ramadori G, Hess G, Hercend T Meyer zum Büschenfelde K-H (1988) Clonal analysis of human lymphocytes infiltrating the liver in chronic hepatitis B and primary biliary cirrhosis. Eur J Immunol 18: 1447–1452
Missale G, Redeker A, Person J, Fowler P, Guilhot S, Schlicht H-J, Ferrari C, Chisari FV (1993) HLA-A31 and Aw68 restricted cytotoxic T cell responses to a single hepatitis B virus nucleocapsid epitope during acute viral hepatitis. J Exp Med 177: 751–762
Monaco, JJ (1992) A molecular model of MHC class I restricted antigen processing. Immunol Today 13: 173–179
Mosmann TR, Cherwinski H, Bond MW, Giedlin MA, Coffman RL(1986) Two types of murine helper T cell clone. I. Definition according to profiles of lymphokine activities and secreted proteins J Immunol 136: 2348–2357
Nassal M (1992) The arginine-rich domain of the hepatitis B virus core protein is required for pregenome encapsidation and productive viral positive-strand DNA synthesis but not for virus assembly. J Virol 66: 4107–4116
Nayersina R, Fowler P, Guilhot S, Missale G, Cerny A, Schlicht H-J, Vitiello A, Chesnut R, Person JL, Redeker AG, Chisari FV (1993) HLA A2 restricted cytotoxic T lymphocyte responses to hepatitis B surface antigen group and subtype specific epitopes during hepatitis B virus infection. J Immunol 150: 4659–4671
Nebert DW, Nelson DR (1991) P450 gene nomenclature based on evolution. Methods Enzymol 206: 3–11
Okamoto H, Sugiyama Y, Okada S, Kurai K, Akahane Y, Sugai Y, Tanaka T, Sato K, Tsuda F, Miyakawa Y, Mayumi M (1992) Typing hepatitis C virus by polymerase chain reaction with type-specific primers: applications to clinical surveys and tracing infectious sources. J Gen Virol 73: 673–679
Penna A, Chisari FV, Bertoletti A, Missale G, Fowler P, Giuberti T, Fiaccadori F, Ferrai C (1991) Cytotoxic T lymphocytes recognize an HLA-A2 restricted epitope within the hepatitis B virus nucleocapsid antigen. J Exp Med 174: 1565–1570
Phillips RE, Rowland-Jones S, Nixon DF, Gotch FM, Edwards JP, Ogunlesi AO, Rothbard JA, Bangham CRM, Rizza CR, McMichael AJ (1991) Human immunodeficiency virus genetic variation that can escape cytotoxic T cell recognition. Nature 354: 453–459
Pircher H, Moskophidis D, Rohrer U, Bürki K, Hengartner H, Zinkernagel RM (1990) Viral escape by selection of cytotixic T-cell-resistant virus variants in vivo. Nature 346: 629–633
Riddell SR, Rabin M, Geballe AP, Britt WJ, Greenberg PD (1991) Class I MHC-restricted cytotoxic T lymphocyte recognition of cells infected with human cytomegalovirus does not require endogenous viral gene expression. J Immunol 146: 2795–2804
Rothbard JB, Gefter ML (1991) Interactions between immunogeneic peptides and MHC proteins. Annu Rev Immunol 9: 527–565
Rötzschke O, Falk K (1991) Naturally occurring peptide antigens derived from the MHC class I restricted processing pathway. Immunol Today 12: 447–455
Rötzschke O, Falk K, Stevanovic S, Jung G, Rammensee H-G (1992) Peptide motifs of closely related HLA class I molecules encompass substantial differences. Eur J Immunol 22: 2453–2456
Schlicht HJ, Schaller H (1989) The secretory core protein of human hepatitis B virus is expressed on the cell surface. J Virol 63: 5399–5404
Shirai M, Akatsuka T, Pendleton CD, Houghten R, Wychowski C, Mihalik K, Feinstone S, Berzofsky JA (1992) Induction of cytotoxic T cells to a cross-reactive epitope in the hepatitis C virus nonstructural RNA polymerase-like protein. J Virol 66: 4098–4106

Tigges MA, Koelle D, Hartog K, Sekulovich RE, Corey L, Burke RL (1992) Human $CD8^+$ herpes simplex specific cytotoxic T lymphocyte clones recognize diverse virion protein antigens. J Virol 66: 1622–1634

Tiollais P, Charnay P, Vyas GN (1981) Biology of hepatitis B virus. Science 213: 406–411

Walker BD, Flexner C, Birch-Limberger K, Fisher L, Paradis TJ, Aldovini A, Young R, Moss B, Schooley RT (1989) Long-term culture and fine specificity of human cytotoxic T-lymphocyte clones reactive with human immunodeficiency virus type 1. Proc Natl Acad Sci USA 86: 9514–9518

Yang P-M, Su I-J, Lai M-Y, Huang G-T, Hsu H-C, Chen D-S, Sung J-L. (1988) Immunohistochemical studies on introhepatic lymphocyte infiltrates in chronic type B hepatitis, with special emphasis on the activation status of the lymphocytes. Am J Gastroenterology 83: 948–953

Yasukawa M, Zarling JM (1984) Human cytotoxic T cell clones directed against herpes simplex virus-infected cells. I. Lysis restricted by HLA class II MB and DR antigens. J Immunol 133: 422–427

Cytotoxic T Lymphocytes in Humans Exposed to *Plasmodium falciparum* by Immunization or Natural Exposure*

S. L. HOFFMAN[1], M. SEDEGAH[1,2], and A. MALIK[1]

1 The Problem of Malaria and Efforts to Control the Disease

Malaria is one of the most important infectious diseases in the world. It is estimated that 2.1 billion people live in areas of the world where malaria is transmitted and that there are 100–300 million new cases of malaria and one

[1] Malaria Program, Naval Medical Research Institute, Bethesda, MD 20889-5607, USA
[2] Pan American Health Organization, Washington DC 20037, USA

* The opinions and assertions herein are the private ones of the authors and are not to be construed as official or as reflecting the views of the U.S. Navy or the Naval Service at large.

Current Topics in Microbiology and Immunology, Vol. 189

to two million deaths caused by malaria every year (WHO 1991). In the past 10–20 years, the severity of the malaria problem has worsened in many areas because of the emergence of drug-resistant strains of the parasite, resistance to insecticides of the *Anopheles* sp. mosquitoes that transmit the disease, socioeconomic problems that have led to a decreased capacity to optimally utilize existing tools to combat the disease, and movement of nonimmune populations into areas where malaria is transmitted. Accordingly, there are now renewed efforts to control this disease. A major focus of research efforts is to develop vaccines against malaria, including vaccines designed to produce protective cytotoxic T lymphocytes (CTL; HOFFMAN et al. 1991). To understand how such vaccines might work, it is important to understand the life cycle of the parasite.

2 Life Cycle of *Plasmodium falciparum*

Plasmodium falciparum is transmitted to humans by the bite of female *Anopheles* sp. mosquitoes. The mosquito injects sporozoites that pass to the liver, where they enter hepatocytes within 30 min (FAIRLEY 1947). It is unknown whether the sporozoites pass through another cell, such as a reticuloendothelial cell, on the way from the circulation to hepatocytes. Most of the uninucleate sporozoites develop in hepatocytes during 6–10 days to mature liver stage schizonts, which have 10000–40000 uninucleate merozoites. After rupturing out of hepatocytes, each merozoite can invade an erythrocyte, where during the next 48 h each uninucleate merozoite can develop to a mature erythrocytic stage schizont with an average of 16 merozoites. These erythrocytic stage schizonts then rupture, thus releasing merozoites, each of which can then reinvade an erythrocyte, initiating the cycle of amplification, rupture, and reinvasion that leads to increasing levels of parasitemia and the clinical manifestations of malaria. Some of the parasites within infected erythrocytes do not develop to erythrocytic stage schizonts, but instead become sexual forms of the parasite called gametocytes. If a mosquito ingests blood infected with gametocytes, the gametocytes develop during an average of 14 days to sporozoites that can be inoculated into other humans. Parasites that cause malaria in rodents such as *P. yoelii* and *P. berghei* have a similar life cycle, but develop much more rapidly in the liver (40–48 h). Since sporozoites and merozoites in circulation are extracellular and mature human erythrocytes are not known to have major histocompatibility antigens on their surface, infected hepatocytes are considered the primary, and perhaps the only, target of CTL against *P. falciparum*.

3 Rationale for Work on $CD8^+$ CTL Against Pre-erythrocytic *Plasmodium* Species Antigens

Immunization of mice (NUSSENZWEIG et al. 1969) and humans (CLYDE et al. 1973; RIECKMANN et al. 1979) with radiation attenuated *Plasmodium* sp. sporozoites protects against challenge with live sporozoites, but does not protect against challenge with infected erythrocytes. The protective immunity induced by the irradiated sporozoite vaccine must therefore be directed against extracellular sporozoites in the circulation (antibodies) or against parasites within hepatocytes (antibodies or cellular immune responses). This is consistent with the observation that irradiated sporozoites develop only partially within hepatocytes and never mature to the late liver schizont stage. Because the sera from mice (VANDERBERG et al. 1969) and humans (CLYDE et al. 1973) immunized with irradiated sporozoites precipitates the surface coat of sporozoites, for many years antibodies against sporozoites were thought to be the major immune effectors responsible for this protection. However, in 1977 it was reported that mu-suppressed mice (mice that could not produce antibodies) immunized with irradiated sporozoites were protected against challenge (CHEN et al. 1977), and several years ago it was shown that adoptive transfer of T cells from mice immunized with irradiated sporozoites protected naive mice against challenge with live sporozoites at a time when the mice had no detectable circulating antisporozoite antibodies (EGAN et al. 1987). Furthermore, in some strains of mice the immunity induced by immunization with irradiated sporozoites was completely reversed by in vivo depletion of $CD8^+$ T lymphocytes (SCHOFIELD et al. 1987; WEISS et al. 1988). The antibodies induced by this form of immunization were not adequate to protect on their own, and it appeared that $CD8^+$ CTL were required for this protective immunity. CTL must recognize their target peptides in combination with major histocompatibility complex (MHC) proteins. These data indicated that the protective CTL could not be directed against extracellular sporozoites in the circulation and must be directed against *Plasmodium* sp. peptides in association with class I MHC molecules on the surface of infected hepatocytes. However, when these discoveries were made, no *Plasmodium* sp. antigen had ever been detected on the surface of infected hepatocytes, inflammatory infiltrates had been shown to be infrequent in the livers of mice infected with malaria (MEIS et al. 1987), and there was considerable debate regarding whether hepatocytes express class I MHC molecules and whether immune CTL could pass from the sinusoids into the space of Disse and attack infected hepatocytes. Within a year it was shown that mice immunized with irradiated *P. berghei* sporozoites and challenged with live sporozoites developed antigen-specific, $CD8^+$ T cell-dependent infiltrates in their livers and that spleen cells from these mice eliminated *P. berghei*-infected hepatocytes from in vitro culture in an MHC-restricted and antigen-specific manner (HOFFMAN et al. 1989a). These studies demonstrated that *Plasmodium* sp. peptides were recognized on infected hepatocytes by T cells and indicated that immune cells could reach, recognize, and destroy infected hepatocytes. They did not indicate which specific *Plasmodium* sp. antigens on hepatocytes were recognized by CTL.

Another intriguing observation supporting the role of CTL in protection against malaria was made several years later. An association was noted between the presence of HLA Bw53 (class I) and protection against severe malaria in Gambians (HILL et al. 1991). HLA Bw53 was found in 16.9% of cases of severe malaria, in 25.4% of controls with mild malaria, and in 25% of adults without malaria. The relative risk of severe malaria among individuals with HLA Bw53 compared with those without this allele was 0.59.

HLA Bw53 is found in 15%–40% of the population of sub-Saharan Africa, but is found in less than 1% of Caucasians and Orientals; like the sickle cell trait, it may have been selected because it protects against severe malaria. The presence of HLA Bw53 is not as protective as the HbS carrier state; only 1.2% of patients with severe malaria in the Gambia study had HbS, while 12.9% of controls with mild malaria carried HbS (relative risk 0.08). Nonetheless, the decreased association of HLA Bw53 with severe malaria suggests that naturally acquired CTL against *Plasmodium* sp. proteins protect against severe malaria. The logical targets for CTL are infected hepatocytes. Most deaths occur in children less than 5 years of age, and these children are not protected against developing malaria infections. Thus, such data raise the possibility that under natural conditions of exposure, CTL against infected hepatocytes reduce mortality by reducing the burden of infection, not by preventing infection. This is consistent with the recent report that use of insecticide-impregnated bednets does not reduce the incidence of parasitemia in Gambian villages, but has a dramatic effect on mortality (ALONSO et al. 1991).

4 $CD8^+$ CTL Against the Circumsporozoite Protein

4.1 Rationale for Work

When work began in earnest on CTL in malaria, the only characterized sporozoite or liver (pre-erythrocytic) stage antigen available for study was the circumsporozoite (CS) protein (YOSHIDA et al. 1980). Kumar reported that B10.BR mice immunized with irradiated *P. falciparum* sporozoites and recombinant vaccinia expressing the *P. falciparum* CS protein (PfCSP) produced CTL against a single 23-amino acid region from the carboxy terminus of the CS protein, Pf 7G8 CS368–390 (KPKDELDYENDIEKKICK-MEKCS; KUMAR et al. 1988). It was next reported that BALB/c mice immunized with *P. berghei* (ROMERO et al. 1989) or *P. yoelii* (WEISS et al. 1990) sporozoites produced CTL against 12-amino acid (residues 249–260, NDDSYIPSAEKI) and 16-amino acid (residues 281–296, SYVPSAEQILEFVKQI) regions of the respective CS proteins. WEISS demonstrated that the CTL eliminated infected hepatocytes from in vitro culture; the CTL recognized this 16-amino acid peptide on the surface of infected hepatocytes with class I MHC. ROMERO reported that adoptive transfer of CTL clones against the analogous epitope on the *P. berghei* CS protein completely protected against challenge in vivo. RODRIGUES (RODRIGUES et al. 1991) subsequently reported that adoptive transfer of CTL clones against the *P. yoelii* CS protein protected even if transferred 3 h after sporozoites were inoculated, at a time when the sporozoites had left the circulation and were within hepatocytes. Infected hepatocytes were clearly the target of these CTL. RODRIGUES also showed that protective clones could be found in the liver in apposition to infected hepatocytes (RODRIGUES et al. 1992), but that only CTL clones expressing the adhesion molecule CD44 were protective. This suggests that in addition to epitope specificity and cytolytic capacity, homing is critical for protection. This view has been supported by studies in the *P. yoelii* and *P. berghei* rodent model system that show that the induction of CTL by active immunization does not indicate that an animal will be protected (FLYNN et al. 1990; SEDEGAH et al. 1992a; SATCHIDANANDAM et al. 1991).

The mechanism by which the CTL eliminate infected hepatocytes has as yet not been established. Schofield showed that treatment of fully immune A/J mice with antibodies to γ-interferon abrogated the irradiated sporozoite-induced protective immunity (SCHOFIELD et al. 1987). This was not the case in BALB/c mice immunized with irradiated *P. berghei* sporozoites (HOFFMAN et al. 1989a). However, Weiss has recently shown that protective immunity conferred by adoptive transfer of a $CD8^+$CTL clone against a 10-amino acid peptide (SYVPSAEQIL) on the *P. yoelii* CS protein is eliminated by in vivo treatment with anti-γ-interferon (WEISS et al. 1992). Weiss has also shown that this $CD8^+$ CTL clone protects against both *P. yoelii* and *P. berghei*, despite the fact that the two CS proteins vary by two amino acids at the site of the CTL epitope (WEISS et al. 1992).

In addition to the adoptive transfer experiments, it has been shown that immunization of BALB/c mice with recombinant *Salmonella typhimurium* expressing the *P. berghei* CS protein induces protective immunity in 50%–75% of mice and that this protection is eliminated by in vivo depletion of $CD8^+$ T cells (AGGARWAL et al. 1991). Likewise, immunization of BALB/c with recombinant P815 mastocytoma cells expressing the *P. yoelii* CS protein also induces $CD8^+$ T cell-dependent protective immunity in 50%–75% of recipient mice (KHUSMITH et al. 1991).

4.2 CD8$^+$ T Cell-Dependent Cytolytic Activity Against the *P. falciparum* Circumsporozoite Protein

4.2.1 Volunteers Immunized with Radiation-Attenuated Sporozoites

The studies in the rodent model systems provided a compelling rationale for developing vaccines for humans that would induce similar CTL against the PfCSP. However, at the time such work was contemplated, no one had shown that humans produced CTL to any *Plasmodium* sp. protein, and in fact, there was only one report in the world's literature demonstrating that humans produced CTL against any parasite protein (YANO et al. 1989). MALIK and collaborators (1991) developed an assay for demonstrating CD8$^+$ CTL against the PfCSP. To provide effector cells, humans were immunized by the bites of previously irradiated *Anopheles stephensi* mosquitoes carrying NF54/3D7 *P. falciparum* sporozoites in their salivary glands (EGAN et al. 1993). Peripheral blood mononuclear cells were harvested and stimulated in vitro with recombinant vaccinia virus expressing the PfCSP (7G8 clone of *P. falciparum*) for 6 days. These effectors were then incubated with autologous Epstein-Barr virus (EBV)-transformed B lymphocytes that had been transiently transfected with the gene encoding the 7G8 PfCSP 48 h previously. Using this method, MALIK and colleagues demonstrated that three of four immunized volunteers produced CTL against the PfCSP and that this cytolytic activity was antigen specific, genetically restricted, and dependent on CD8$^+$ T lymphocytes. They reported that cytolytic activity could not be demonstrated each time an assay was run; six of nine, seven of 11, and two of 12 assays were positive in the three individuals shown to have cytolytic activity. They also reported the presence of cytolytic activity 13 weeks after last immunization in one volunteer.

Stimulating with vaccinia-infected cells in vitro was consumptive of large numbers of human cells. Since a peptide including amino acids 368–390 of the Pf 7G8 CS (Table 1) had previously been shown to include a CTL epitope in B10.BR mice (KUMAR et al. 1988), this peptide was studied in the same system. Three of the four volunteers were shown to have CTL directed at one or more epitopes included within this peptide. In one volunteer, overlapping 20-amino acid peptides representing the entire carboxy terminus of the 7G8 CS protein were studied, and only peptide 368–390 was able to label target cells for killing by CTL. CTL from this volunteer recognized both the 7G8 *P. falciparum* sequence and the sequence based on the immunizing sporozoites (3D7) that differed by one amino acid (Table 1). It has not been established whether the variant amino acid residue is actually included within the CTL epitope. However, this finding is of potential importance because of concerns regarding polymorphism of this epitope and the reasons why polymorphism has developed at this site. If the polymorphism is the result of immune selection and CTL against one sequence do not recognize CTL against variant

Table 1. Variation in the region of the only know human $CD8^+$ CTL epitope on the *P. falciparum* circumsporozoite (CS) protein

Strain or isolate	Amino acid residue										
	367	368	369	370	371	372	373	374	375	376	377
7G8 T9/101;406_{3-5}; T4R	**N**	K	**P**	**K**	D	**E**	L	**D**	Y	**E**	**N**
NF54/3D7; $XP_{8,9}$; $X_{10,11}$; T9-98; $406_{1,7,8}$; 427_{1-10}; 419_{10}; GAM1										A	
AE7, PNG2						Q					
ItG2: W/L; PNG1, PNG X 22; T9-94; WEL	D					Q					
MCK^+; AE28; GAM2; MS2; Palo Alto; $366_{1,5,8-10}$ 399_{1-10}; LE5						Q				A	
HB3; X5; $XP_{12,13}$; PNG4; BRA1	G		S								
PNG3	D					Q		C			S
GAM3; $366_{2-4,6,7}$; 406_{10}; 419_{1-9}						Q		N			
GAM4; 406_2				R						A	
GAM5; $406_{3,9}$										A	D

In mice and humans, a peptide including amino acid residues 368–390 of the 7G8 *P. falciparum* CS protein (Pf 7G8 CS 368–390, KPKDELDYENDIEKKICKMEKCS; DAME et al. 1984) has been shown to include a $CD8^+$ CTL epitope. No variation has been identified from amino acids 378–390. Amino acids 367–377 of the CS protein of the 7G8 clone of *P. falciparum* are shown, and identified variations in this region (LOCKYER et al. 1989; DOOLAN et al. 1992; SHI et al. 1992) are listed. Amino acid 367 is also included, because there are data suggesting that there is a CTL epitope in the region from amino acid 351–370 (DOOLAN et al. 1991), and no variation has been shown from amino acid residues 351–366, but variation has been identified at amino acid residue 367.

sequences, the usefulness of a vaccine designed to produce CTL against only one variant will be limited. The importance of polymorphism at important $CD8^+$ CTL sites remains to be elucidated. However, variation has been identified in isolates from around the world (Table 1; LOCKYER et al. 1989; DOOLAN et al. 1992; SHI et al. 1992), suggesting that the variation is the result of immune selection (GOOD et al. 1988).

Having established that volunteers immunized with irradiated *P. falciparum* sporozoites produced CTL against the PfCSP, the investigators asked whether the demonstration of CTL by their assay indicated that a volunteer would be protected. As predicted by previous murine studies (FLYNN et al. 1990; SEDEGAH et al. 1992a; SATCHIDANANDAM et al. 1991), the presence of CTL against this epitope did not indicate that a volunteer would be protected aginst challenge, and the absence of cytolytic activity did not indicate that a volunteer would not be protected (MALIK et al. 1991).

Table 2. HLA types of the four Kenyans naturally exposed to malaria and three individuals immunized with irradiated *P. falciparum* sporozoites who have been shown to have $CD8^+$ T cell dependent cytolytic activity against peptide Pf CS 368–390 (MALIK et al. 1991; SEDEGAH et al. 1992)

Volunteer	Class I				Class II			CTL
	A	B	Bw4/w6	C	DR	DRw52/53	Dq	
6	2, 32	w48, 45	w6, w6	w6	w8, w11	w52	w7,w1	+
9	30, w34	w42, w48	w6, w6	w4	ND	ND	ND	+
10	2, 28	51, 15	w4, w4	w4	w11, 7	w52, w53	w7, w2	+
8	30, 32	w58, -	w4	w6	w11, -	w52	w1, w7	+
1H	11, 24	35, -	ND	w4, -	[a]	[a]	w1	+
3W	1, 3	8, 35	ND	w4,	1, 7	w53, -	w1, w2	+
4R	1, 28	44, w57	ND	w6, -	7, w11	w52, w53	w3	+

[a] Not typeable by conventional methods.

4.2.2 Kenyans Naturally Exposed to Malaria

Knowing that humans immunized with irradiated sporozoites produced $CD8^+$ CTL against the region of PfCSP including amino acids Pf 7G8 CS 368–390, Sedegah and coworkers set out to determine whether humans naturally exposed to malaria produced CTL against this region of the CS protein. This region was of particular interest because it had been previously shown that there was a correlation between lymphocyte proliferation to peptides Pf 7G8 CS 361–380 and 371–390 and resistence to reinfection with malaria among Kenyan adults (HOFFMAN et al. 1989; MASON et al., unpublished). Eleven Kenyans were selected because their lymphocytes had been shown to proliferate after stimulation with one of these peptides and because they were also resistant to reinfection with malaria (nine of the 11 volunteers). They were studied at a single time point at the end of a period of low malaria transmission, and four of the 11 volunteers were shown to have cytolytic activity against autologous EBV-transformed B cells pulsed with the Pf 7G8 CS 368–390 peptide. In three of the four cases, the cytolytic activity was reversed by depletion of $CD8^+$ T cells, but in no case did depletion of $CD4^+$ T cells reduce cytolytic activity (SEDEGAH et al. 1992b). Individuals of various HLA phenotypes have been shown to have $CD8^+$ T cell-dependent cytolytic activity against target cells pulsed with peptide 368–390 (Table 2). Thus far, minimal CTL epitopes have not been identified and T cell clones against this region of the CS protein have not been produced. Thus, restriction elements have not been clearly delineated.

4.2.3 HLA B35-Restricted Activity Among Gambians Naturally Exposed to Malaria

Using the techniques described in Sect. 5.1 below, HILL and colleagues (1992) eluted peptides from HLA B35. In addition to a proline at position 2, they found these peptides generally included a tyrosine at position 9. They found ten peptide sequences from the four pre-erythrocytic stage proteins described below that had prolines at positions 2 and tyrosines at position 8 (six sequences), position 9 (one sequence), or position 10 (three sequences). Six of these sequences were from PfCSP, and four of the six were from variants of the first eight to ten amino acids of Pf CS 7G8 368–390, the previously identified human CTL epitope (see Sects. 4.2.1 and 4.2.2 above). One of eight Gambians tested had cytolytic activity against peptide Pf CS 7G8 368–375 (KPKDELDY). Cells from this volunteer that were stimulated with peptide KPKDELDY did not lyse targets pulsed with peptides KPKD**Q**L**N**Y or K**S**KDELDY, sequences from other known variants (Table 1). Cells from a second volunteer recognized peptide K**S**KDELDY, but did not recognize the variant peptides KPKDELDY, KPKD**Q**LDY, or KPKD**Q**L**N**Y (Table 1). This cytolytic activity was inhibited by an anti-CD8 monoclonal antibody. These data demonstrate for the first time that single amino acid changes at the site of PfCSP CTL epitopes can eliminate recognition by CTL. Identification of a minimal CTL epitope in the region of Pf CS 7G8 368–390 with the greatest polymorphism supports the hypothesis that this polymorphism has arisen secondary to immune pressure (GOOD et al. 1988).

5 CD8$^+$ CTL Against Liver Stage Antigen-1

5.1 HLA Bw53-Restricted, CD8$^+$ T Cell-Dependent Cytolytic Activity Against Liver Stage Antigen-1 in Gambians Naturally Exposed to Malaria

Elegant studies that followed up the demonstration of an association between HLA Bw53 and lack of severe malaria (see Sect. 3 above) have recently been reported (HILL et al. 1992). The investigators sequenced peptides eluted from HLA B53 and found that most peptides had a proline at position 2. They then synthesized 60 8- to 10-amino acid peptides from four *P. falciparum* pre-erythrocytic stage antigens, Pf CSP (DAME et al. 1984), TRAP/Pf SSP2 (ROBSON et al. 1988; ROGERS et al. 1992a), liver stage antigen-1 (LSA-1; GUERIN-MARCHAND et al. 1987; ZHU and HOLLINGDALE 1991), and Pfs16 (MOELANS et al. 1991), that had prolines in position 2. These peptides were studied to determine whether they bound to HLA B53 in an HLA assembly assay. Eight peptides were found to bind, three from PfCSP, two from TRAP/PfSSP2, one from LSA-1, and two from Pfs16. Six individuals in one village and nine individuals in a village with much lower transmission of malaria were tested during a period of low malaria transmission. Using a classical restimulation CTL assay, they showed that none of the individuals had CTL against the PfCSP, PfSSP2/TRAP, and Pfs16 peptides.However, three of six volunteers in the first village and one of nine volunteers in the second village had HLA B53, peptide-specific, CD8$^+$ T cell-dependent cytolytic activity against the LSA-1 peptide referred to as ls6 (KPIVQYDNF). This epitope was shown to be invariant in nine *P. falciparum* isolates that they sequenced. These findings suggest that CTL against this LSA-1 peptide may be involved in the partial protection against severe malaria associated with HLA Bw53.

5.2 HLA B35-Restricted CTL Against Liver Stage Antigen-1

One of eight Gambians was shown to have cytolytic activity against the LSA-1 peptide (ls8, KPNDKSLY) eluted from HLA B35 as described in Sects. 4.2.3 and 5.1 above (Hill et al. 1992). There was no indication as to whether the cytolytic activity was dependent on $CD8^+$ T cells.

6 $CD4^+$ CTL Against the Circumsporozoite Protein

6.1 Rationale for Work

Del Giudice and colleagues immunized mice subcutaneously at the base of the tail with a peptide including amino acids 59–79 of the *P. yoelii* CS protein (YNRNIVNRLLGDALNGKPEEK) and using regional lymph node cells produced a $CD4^+$ T cell clone that recognized this peptide and a shorter peptide (PyCS 61–70 in BALB/c mice and PyCS 59–70 in C57BL/6 mice; Del Giudice et al. 1990). This clone eliminated infected hepatocytes from culture (Renia et al. 1991) and adoptively transferred protection against *P. yoelii*. The clone was never shown to have cytolytic activity against target cells pulsed with the peptide, but the in vitro activity against infected hepatocytes was not reversed by anticytokine treatment. It is not certain whether these clones are directly cytolytic or whether they mediate protection through another mechanism. Nonetheless, working in the less infective, but uniformly lethal, *P. berghei* system, Corradin and colleagues have found that immunization of mice at the base of the tail with the analogous *P. berghei* peptide presented as a branched chain polymer protects mice (unpublished observation). The mechanism of this protection has not been established, but such data identify protective $CD4^+$ T cells against an epitope amino terminal of the repeat region of the *P. yoelii* CS protein.

6.2 $CD4^+$ CTL Against the *P. falciparum* Circumsporozoite Protein in Volunteers Immunized with Irradiated *P. falciparum* Sporozoites

Moreno and colleagues have recently reported the production of a $CD4^+$ T cell clone with cytolytic activity from an individual immunized with radiation

Table 3. Variation in the region of the only known human CD4⁺ CTL epitope on the *P. falciparum* circumsporozoite (cs) protein

Strain or isolate	*Amino acid residue (7G8 sequence)*													
	329	330	331	332	333	334	335	336	337	338	339	340	341	342
7G8	**K**	H	I	**E**	**Q**	Y	L	**K**	**K**	I	**K**	N	S	**I**
NF54/3D7; $XP_{8,9,}$ $X_{10,11,}$; $427_{1-4,6-10}$				K	E			N			Q			L
T4; T4R; T9-101; WEL; W/L; PNG X 22; T9-94; MCK⁺; AE28; AE7; HB3; X_5; $XP_{12,13}$; PNG2; BRAI											Q			L
ItG2G1; ItG2					K						Q			L
366_1; 399_{1-10}; 406_{1-9} 419_{10};	Q				K				T					L
T9-98									T					L
427_5								N	T					L
LE5, 366_9	Q				K				T		Q			L
MS2, Palo Alto Brazil X4; BRA2	Q				K				R		Q			L
366_{2-7}	Q				K				Q		R			L
$366_{8,10}$	Q				K				Q		Q			L
406_{10}; 419_{1-9}	Q				K				Q					L
GAM1	Q				K				T		Q			L
GAM2	Q				K				T					L
GAM3	Q				K			Q			Q			L
GAM4	Q				K			Q						L
GAM5								N	T		Q			L
GAM6	Q				K			Q			R			L

In humans, a peptide including amino acid residues 326–345 of the NF54/3D7 *P. falciparum* CS protein (Pf 3D7 CS 326–345, EYLN**KIQNSLSTEW**SPCSVT; CASPERS et al. 1989; CAMPBELL 1989) that is analogous to amino acid residues 333–352 of the 7G8 CS protein seqence (DAME et al. 1984) has been shown to include a CD4⁺ CTL epitope (MORENO et al. 1991). The minimal epitope recognized in the context of HLA DR7 has been shown to be amino acids Pf 3D7 CS 330–339 (KIQNSLSTEW) (= Pf 7G8 337–346). No variation has been identified in amino acids Pf 3D7 CS 336–345 (Pf 7G8 CS 343–352). Amino acids 329–342 of the 7G8 clone of *P. falciparum* are shown, and identified variations among isolates in the region (LOCKYER et al. 1989; DOOLAN et al. 1992; SHI et al. 1992) are listed. Amino acid residues 329–332 are included, since Pf 7G8 326–345 (PSDKHIEQYLKKIKNSISTE) has been shown to include a T-helper epitope, and no variation has been shown in amino acids 326–328 of the 7G8 sequence.

attenuated *P. falciparum* sporozoites (MORENO et al. 1991). They have mapped the epitope to residues 330–339 (KIQNSLSTEW) of the NF54/3D7 sequence of the PfCSP (CASPERS et al. 1989; CAMPBELL 1989) and demonstrated that this peptide is recognized in the context of HLA DR7. This epitope corresponds to amino acids 337–346 of the 7G8 sequence of the PfCSP (DAME et al. 1984) and is found in a polymorphic region of the CS protein (Table 3; LOCKYER et al. 1989; DOOLAN et al. 1992; SHI et al. 1992). It has therefore been suggested that, as for the $CD8^+$ CTL epitope described above, the variation in this region is a result of immune selection.

7 $CD4^+$ CTL Against Other Pre-erythrocytic Antigens in Murine Models

TSUJI and colleagues (1990) immunized mice with irradiated *P. berghei* sporozoites, stimulated spleen cells with an extract of blood stage *P. berghei* parasites, and derived a $CD4^+$ T cell clone that produces interleukin 2 (IL-2) and γ-interferon when stimulated with blood stage parasite extract. The clone also lyses target cells pulsed with blood stage *P. berghei* extract. More importantly, adoptive transfer of this clone protects against sporozoite challenge, but not against blood stage challenge. This strongly suggests that the clone is recognizing a parasite antigen expressed in infected hepatocytes. Neither the *P. berghei* antigen recognized by this clone, nor its *P. falciparum* analog has been identified.

8 CTL Against the *P. falciparum* Circumsprozoite Protein Among Australians Naturally Exposed to Malaria

Doolan and colleagues stimulated peripheral blood mononuclear cells from Australians who had been previously exposed to malaria (DOOLAN et al. 1991) with a mixture of CS protein peptides. Their studies showed that after two cycles of stimulation, cells from three of 42 donors had cytolytic activity against a mixture of overlapping 20-amino acid peptides spanning amino acid residues 341–412 (Pf CSP 7G8). Like the results with volunteers immunized with irradiated sporozoites (MALIK et al. 1991), cytolytic activity could not be demonstrated each time an assay was run, suggesting to the investigators that there is a low frequency of these cells in the peripheral blood. They did not report on whether the activity was dependent on a particular subset of T lymphocytes or genetically restricted, but two of the three positive donors were HLA Bw57. They produced a T cell line from one donor and demonstrated that the line lysed target cells pulsed with peptides Pf7G8 CS 351–370, 371–390, or 376–395. These data indicate that there is at least one CTL epitope within residues Pf7G8 CS 351–370 and at least one more within residues Pf7G8 CS 371–390. However, the T cell subset specificity and genetic restriction of this cytolytic activity remains to be established.

9 Rationale for Work on CD8$^+$ CTL Against Sporozoite Surface Protein-2

Immunization with irradiated sporozoites completely protects against malaria, but none of the subunit *P. berghei* or *P. yoelii* CS protein vaccines have given protection comparable to the irradiated sporozoite vaccine. Furthermore, in the human studies the presence of CTL against the PfCSP did not guarantee that the individual would be protected and, likewise, one individual who was not shown to have CTL was protected against challenge (MALIK et al. 1991). Considering the complexity of sporozoites, it was not logical to assume that all protection induced by the whole organism vaccine was mediated by CTL against a single short stretch of amino acids on a single protein. Thus, there has been considerable effort to identify additional targets of irradiated sporozoite-induced protective immunity. Mice were immunized with irradiated *P. yoelii* sporozoites, and a monoclonal antibody directed at a 140-kDa sporozoite protein was produced (CHAROENVIT et al. 1987). The gene encoding this protein was cloned and sequenced (HEDSTROM et al. 1990; ROGERS et al. 1992a), and the protein was named sporozoite surface protein-2 (PySSP2). To determine whether immunization with irradiated sporozoites not only produced antibodies, but also CTL against PySSP2, a 1.5-kb fragment of the gene encoding PySSP2 was transfected into P815 mouse mastocytoma cells. When these transfected cells were used as targets in CTL assays, mice immunized with irradiated sporozoites were shown to produce CTL against PySSP2 (KHUSMITH et al. 1991). KHUSMITH et al. subsequently produced CD8$^+$ CTL clones against PySSP2 and showed that adoptive transfer of one of these clones completely protected against challenge, establishing that CTL against PySSP2 could completely protect against this highly infectious parasite in the absence of any other parasite specific immune responses (S. KHUSMITH, unpublished). Like the anti-CS protein CTL clones, these clones protected when adoptively transferred into mice 3 h after sporozoite inoculation, indicating that they were attacking infected hepatocytes. Mice were also immunized with the P815 cells expressing PySSP2, and approximately 50% were protected against challenge (KHUSMITH et al. 1991). KHUSMITH then went on to show that although immunization with the *P. yoelii* CS protein or PySSP2 vaccines gave only partial protection against malaria (50%–75%), immunization with transfected P815 cells expressing, *P. yoelii* CSP and PySSP2 produced 100% protection (KHUSMITH et al. 1991). Furthermore, as after immunization with irradiated sporozoites, this protective immunity was completely reversed by in vivo depletion of $CD8^+$ T cells.

The gene encoding the *P. falciparum* SSP2 (PfSSP2) has now been identified and characterized (ROGERS et al. 1992b) and shown to be the previously described thrombospondin-related anonymous protein (TRAP; ROBSON et al. 1988). Work is in progress to identify CTL epitopes on PfSSP2 and to produce vaccines that will induce protective CTL in humans against PfSSP2 and PfCSP.

10 Conclusion

The study of human CTL against *Plasmodium* sp. peptides is in its infancy. There is a compelling rationale for work on the identification, characterization, and induction of class I-restricted, $CD8^+$ CTL against pre-erythrocytic stages of *P. falciparum*. The data supporting work on $CD4^+$ CTL is not as complete, but is also solid. Work is now, or will soon be, underway in a number of laboratories to use synthetic peptide, recombinant protein, live vector-delivered antigen and plasmid DNA vaccines to induce CTL in humans against PfCSP, PfSSP2, PfLSA-1, and other *P. falciparum* proteins expressed in infected hepatocytes. In parallel studies, a number of scientists are attempting to characterize further the qualities of CTL associated with protection, to identify and characterize additional CTL epitopes on these proteins, and to optimize vaccine delivery systems in the rodent model systems. Duplicating the non-strain-specific, sterile, consistently protective immunity induced in humans by the irradiated sporozoite vaccine will almost certainly require constructing vaccines that induce a essentially all vaccinees protective CTL against at least one epitope that is present in all strains of parasites. It will probably also require induction of protective antibody responses.

Acknowledgements. This work was supported in part by the Naval Medical Research and ONR grant N00014-89-J-1856 to the Pan American Health Organization.

References

Aggarwal A, Kumar S, Jaffe R, Hone D, Gross M, Sadoff J (1991) Oral Salmonella: malaria circumsporozoite recombinants induce specific $CD8^+$ cytotoxic T cells. J Exp Med 172: 1083–1090

Alonso PL, Lindsay SW, Armstrong JR, Conteh M, Hill AG, David PH, Fegan G (1991) The effect of insecticide-treated bed nets on mortality of Gambian children. Lancet 337: 1499–1502

Campbell JR (1989) DNA sequence of the gene encoding a *Plasmodium falciparum* malaria candidate vaccine antigen. Nucleic Acids Res 17: 5854

Caspers P, Gentz R, Matile H, Pink JR, Sinigaglia F (1989) The circumsporozoite protein gene from NF54, a *Plasmodium falciparum* isolate used in malaria vaccine trial. Mol Biochem Parasitol 35: 185–190

Charoenvit Y, Leef ML, Yuan LF, Sedegah M, Beaudoin RL (1987) Characterization of *Plasmodium yoelii* monoclonal antibodies directed against stage-specific antigens. Infect Immun 55: 604–608

Chen DH, Tigelaar RE, Weinbaum FI (1977) Immunity to sporozoite-induced malaria infection in mice. 1. The effect of immunization of T and B cell-deficient mice. J Immunol 118: 1322–1327

Clyde DF, McCarthy VC, Miller RM, Hornick RB (1973) Specificity of protection of man immunized against sporozoite-induced falciparum malaria. Am J Med Sci 266: 398–401

Dame JB, Williams JL, McCutchan TF, Weber JL, Wirtz RA, Hockmeyer WT, Sanders GS, Reddy EP, Maloy WL, Haynes JD, Schneider I, Roberts D, Diggs CL, Miller LH (1984) Structure of the gene encoding the immunodominant surface antigen on the sporozoite of the human malaria parasite *Plasmodium falciparum*. Science 225: 593–599

Del Giudice G, Grillot D, Renia L, Muller I, Corradin G, Louis JA, Mazier D, Lamberd PH (1990) Peptide-primed $CD4^+$ cells and malaria sporozoites. Immunol Lett 25: 59–64

Doolan DL, Houghten RA, Good MF (1991) Location of human cytotoxic T cell epitopes within a polymorphic domain of the *Plasmodium falciparum* circumsporozoite protein. Int Immunol 3: 511–506

Doolan DL, Saul AJ, Good MF (1992) Geographically restricted heterogeneity of the *Plasmodium falciparum* circumsporozoite protein; relevance for vaccine development. Infect Immun 60: 675–682

Egan JE, Weber JL, Ballou WR, Hollingdale MR, Majarian WR, Gorden DM, Maloy WE, Hoffman SL, Wirtz RA, Schneider I, Woollett GR, Young JF and Hockmeyer WT (1987) Efficacy of murine malaria sporozoite vaccines: implications for human vaccine development. Science 236: 453–456

Egan JE, Hoffman SL, Haynes JD, Sadoff JC, Schneider I, Grau GE, Hollingdale MR, Ballou WR, Gordon DM (1993) Humoral immune response in volunteers immunized with irradiated *Plasmodium falciparum* sporozoites. Am J Trop Med Hyg (in press)

Fairley NH (1947) Sidelights on malaria in man obtained by subinoculation experiments. Trans R Soc Trop Med Hyg 40: 621–676

Flynn JL, Weiss WR, Norris KA, Seifert HS, Kumar S, So M (1990) Generation of a cytotoxic T-lymphocyte response using a Salmonella antigen-delivery system. Mol Microbiol 4: 2111–2118

Good MF, Kumar S, Miller LH (1988) The real difficulties for malaria sporozoite vaccine development: nonresponsiveness and antigenic variation. Immun Today 9: 351–355

Guerin-Marchand C, Druilhe P, Galey B, Londono A, Patarapotikul J, Beaudoin RL, Dubeaux C, Tartar A, Mercereau-Puijalon O, Langsley G (1987) A liver-stage-specific antigen of *Plasmodium falciparum* characterized by gene cloning. Nature 329: 164–167

Hedstrom RC, Campbell JR, Leef ML, Charoenvit Y, Carter M, Sedegah M, Beaudoin RL, Hoffman SL (1990) A malaria sporozoite surface antigen distinct from the circumsporozoite protein. Bull WHO 68: 152–157

Hill AV, Allsopp CE, Kwiatkowski D, Anstey NM, Twumasi P, Rowe PA, Bennett S, Brewster D, McMichael AJ, Greenweed BM (1991) Common west African HLA antigens are associated with protection from severe malaria. Nature 352: 595–600

Hill AV, Elvin J, Willis AC, Aidoo M, Allsopp CE, Gotch FM, Gao XM, Takiguchi M, Greenwood BM, Townsand AR, McMichael AJ, and Whittle HC (1992) Molecular analysis of the association of HLA-B53 and resistance to severe malaria. Nature 360: 434–439

Hoffman SL, Isenbarger D, Long GW, Sedegah M, Szarfman A, Waters L, Hollingdale MR, van der Miede PH, Finbloom DS, Ballou WR (1989a) Sporozoite vaccine induces genetically restricted T cell elimination of malaria from hepatocytes. Science 244: 1078–1081

Hoffman SL, Oster CN, Mason C, Beier JC, Sherwood JA, Ballou WR, Mugambi M, Chulay JD (1989b) Human lymphocyte proliferative response to a sporozoite T cell epitope correlates with resistance to falciparum malaria. J Immunol 142: 1299–1303

Hoffman SL, Nussenzweig V, Sadoff JC, Nussenzweig RS (1991) Progress toward malaria preerythrocytic vaccines. Science 252: 520–521

Khusmith S, Charoenvit Y, Kumar S, Sedegah M, Beaudoin RL, Hoffman SL (1991) Protection against malaria by vaccination with sporozoite surface protein 2 plus CS protein. Science 252: 715–718

Kumar S, Miller LH, Quakyi IA, Keister DB, Houghten RA, Maloy WL, Moss B, Berzofsky JA, Good MF (1988) Cytotoxic T cells specific for the circumsporozoite protein of *Plasmodium falciparum*. Nature 334: 258–260

Lockyer MJ, Marsh K, Newbold CI (1989) Wild isolates of *Plasmodium falciparum* show extensive polymorphism in T cell epitopes of the circumsporozoite protein. Mol Biochem Parasitol 37: 275–280

Malik A, Egan JE, Houghton RA, Sadoff JC, Hoffman SL (1991) Human cytotoxic T lymphocytes against *Plasmodium falciparum* circumsporozoite protein. Proc Natl Acad Sci USA 88: 3300–3304

Meis JFGM, Jap PHK, Verhave JP, Meuwissen JHET (1987) Cellular response against exoerythrocytic forms of *Plasmodium berghei* in rats. Am J Trop Med Hyg 30: 506–510

Moelans IIMD, Meiss JFGM, Kocken C, Konongs RNH, Schoemakers JGG (1991) A novel protein antigen of the malaria parasite *Plasmodium falciparum*, located on the surface of gametes and sporozoites. Mol Biochem Parasitol 45: 193–204

Moreno A, Clavijo P, Edelman R, Davis J, Sztein M, Herrington D, Nardin E (1991) Cytotoxic $CD4^+$ T cells from a sporozoite-immunized volunteer recognize the *Plasmodium falciparum* CS protein. Int Immunol 3: 997–1003

Nussenzweig R, Vanderberg J, Most H (1969) Protective immunity produced by the injection of x-irradiated sporozoites of *Plasmodium berghei*: IV. Dose response, specificity and humoral immunity. Milit Med 134: 1176–1182

Renia L, Marussig MS, Grillot D, Pied S, Corradin G, Miltgen F, Del Giudice G, Mazier D (1991) In vitro activity of $CD4^+$ and $CD8^+$ T lymphocytes from mice immunized with a synthetic malaria peptide. Proc Natl Acad Sci USA 88: 7963–7967

Rieckmann KH, Beaudoin RL, Cassells JS, Sell DW (1979) Use of attenuated sporozoites in the immunization of human volunteers against falciparum malaria. Bull WHO 57: 261–265

Robson KJH, Hall JRS, Jennings MW, Harris TJR, Marsh K, Newbold CI, Tate VE, Weatherall DJ (1988) A highly conserved amino-acid sequence in thrombospondin, properdin and in proteins from sporozoites and blood stages of a human malaria parasite. Nature 335: 79–82

Rodrigues MM, Cordey A-S, Arreaza G, Corradin G, Romero P, Maryanski JL, Nussenzweig RS, Zavala F (1991) $CD8^+$ cytolytic T cell clones derived against the *Plasmodium yoelii* circumsporozoite protein protect against malaria. Int Immunol 3: 579–586

Rodrigues MM, Nussenzweig RS, Romero P, Zavala F (1992) The in vivo cytotoxic activity of $CD8^+$ T cell clones correlates with their levels of expression of adhesion molecules. J Exp Med 175: 895–905

Rogers WO, Malik A, Mellouk S, Nakamura K, Rogers MD, Szarfman A, Gordon DM, Nussler AK, Aikawa M, Hoffman SL (1992a) Characterization of *Plasmodium falciparum* sporozoite surface protein 2. Proc Natl Acad Sci USA 89: 9176–9180

Rogers WO, Rogers MD, Hedstrom RC, Hoffman SL (1992b) Characterization of the gene encoding sporozoite surface protein 2, a protective *Plasmodium yoelii* sporozoite antigen. Mol Biochem Parasitol 53: 45–52

Romero P, Maryanski JL, Corradin G, Nussenzweig RS, Nussenzweig V, Zavala F (1989) Cloned cytotoxic T cells recognize an epitope in the circumsporozoite protein and protect against malaria. Nature 341: 323–325

Satchidanandam V, Zavala F, Moss B (1991) Studies using a recombinant vaccinia virus expressing the circumsporozoite protein of *Plasmodium berghei*. Mol Biochem Parasitol 48: 89–100

Schofield L, Villaquiran J, Ferreira A, Schellekens H, Nussenzweig RS, Nussenzweig V (1987) Gamma-interferon, $CD8^+$ T cells and antibodies required for immunity to malaria sporozoites. Nature 330: 664–666

Sedegah M, Chiang CH, Weiss WR, Mellouk S, Cochran MD, Houghton RA, Beaudoin RL, Smith D, Hoffman SL (1992a) Recombinant pseudorabies virus carrying a plasmodium gene: herpesvirus as a new live viral vector for inducing T- and B-cell immunity. Vaccine 10: 578–584

Sedegah M, Sim BKL, Mason C, Nutman T, Malik A, Roberts C, Johnson A, Ochola J, Koech D, Were B, Hoffman SL (1992b) Naturally acquired $CD8^+$ cytotoxic T lymphocytes against *Plasmodium falciparum* circumsporozoite protein. J Immunol 149: 966–971

Shi Y-P, Alpers MP, Povoa MM, Lal AA (1992) Diversity in the immunodominant determinants of the circumsporozoite protein of *Plasmodium falciparum* parasites from malaria-endemic regions of Papua New Guinea and Brazil. Am J Trop Med Hyg 47: 844–851

Sinigaglia F, Guttinger M, Kilgus J, Doron DM, Matile H, Etlinger H, Trezeciak A, Gillessen D and Pink JRL (1988) A malaria T cell epitope recognized in association with most mouse and human MHC class II molecule. Nature 336: 778–780

Tsuji M, Romero P, Nussenzweig RS, Zavala F (1990) $CD4^+$ cytolytic clone confers protection against murine malaria. J. Exp Med 172: 1353–1357

Vanderberg J, Nussenzweig R, Most H (1969) Protective immunity produced by the injection of X-irradiated sporozoites of *Plasmodium berghei.* V. In vitro effects of immune serum on sporozoites. Milit Med 134: 1183–1190

Verhave JP, Strickland GT, Jaffe HA, Ahmed A (1978) Studies on the transfer of protective immunity with lymphoid cells from mice immune to malaria sporozoites. Immunol 121: 1031–1033

Weiss WR, Sedegah M, Beaudoin RL, Miller LH, Good MF (1988) $CD8^+$ T cells (cytotoxic/suppressors) are required for protection in mice immunized with malaria sporozoites. Proc Natl Acad Sci USA 85: 573–576

Weiss WR, Mellouk S, Houghten RA, Sedegah M, Kumar S, Good MF, Berzofsky JA, Miller LH, Hoffman SL (1990) Cytotoxic T cells recognize a peptide from the circumsporozoite protein on malaria-infected hepatocytes. J Exp Med 171: 763–773

Weiss WR, Berzovsky JA, Houghten R, Sedegah M, Hollingdale M, Hoffman SL (1992) A T cell clone directed at the circumsporozoite protein which protects mice against both *P. yoelii* and *P. berghei*. J Immunol (in press)

WHO (1991) World Health Organization Expert Committee on Biological Standardization. Techn Rep Ser 610: 1977

Yano A, Aosai F, Ohta M, Hasekura H, Sugane K, Hayashi S (1989) Antigen presentation by *Toxoplasma gondii*-infected cells to $CD4^+$ proliferative T cells and $CD8^+$ cytotoxic cells. J Parasitol 75: 411–416

Yoshida N, Nussenzweig RS, Ptocynjak P, Nussenzweig V, Aikawa M (1980) Hybridoma produces protective antibodies directed against the sporozoite stage of malaria parasites. Science 207: 71–73

Zhu J, Hollingdale MR (1991) Structure of *Plasmodium falciparum* liver stage antigen-1. Mol Biochem Parasitol 48: 223–226

Subject Index

Current Topics in Microbiology and Immunology

Volumes published since 1989 (and still available)

Vol. 146: **Mestecky, Jiri; McGhee, Jerry (Ed.):** New Strategies for Oral Immunization. International Symposium at the University of Alabama at Birmingham and Molecular Engineering Associates, Inc. Birmingham, AL, USA, March 21–22, 1988. 1989. 22 figs. IX, 237 pp. ISBN 3-540-50841-4

Vol. 147: **Vogt, Peter K. (Ed.):** Oncogenes. Selected Reviews. 1989. 8 figs. VII, 172 pp. ISBN 3-540-51050-8

Vol. 148: **Vogt, Peter K. (Ed.):** Oncogenes and Retroviruses. Selected Reviews. 1989. XII, 134 pp. ISBN 3-540-51051-6

Vol. 149: **Shen-Ong, Grace L. C.; Potter, Michael; Copeland, Neal G. (Ed.):** Mechanisms in Myeloid Tumorigenesis. Workshop at the National Cancer Institute, National Institutes of Health, Bethesda, MD, USA, March 22, 1988. 1989. 42 figs. X, 172 pp. ISBN 3-540-50968-2

Vol. 150: **Jann, Klaus; Jann, Barbara (Ed.):** Bacterial Capsules. 1989. 33 figs. XII, 176 pp. ISBN 3-540-51049-4

Vol. 151: **Jann, Klaus; Jann, Barbara (Ed.):** Bacterial Adhesins. 1990. 23 figs. XII, 192 pp. ISBN 3-540-51052-4

Vol. 152: **Bosma, Melvin J.; Phillips, Robert A.; Schuler, Walter (Ed.):** The Scid Mouse. Characterization and Potential Uses. EMBO Workshop held at the Basel Institute for Immunology, Basel, Switzerland, February 20–22, 1989. 1989. 72 figs. XII, 263 pp. ISBN 3-540-51512-7

Vol. 153: **Lambris, John D. (Ed.):** The Third Component of Complement. Chemistry and Biology. 1989. 38 figs. X, 251 pp. ISBN 3-540-51513-5

Vol. 154: **McDougall, James K. (Ed.):** Cytomegaloviruses. 1990. 58 figs. IX, 286 pp. ISBN 3-540-51514-3

Vol. 155: **Kaufmann, Stefan H. E. (Ed.):** T-Cell Paradigms in Parasitic and Bacterial Infections. 1990. 24 figs. IX, 162 pp. ISBN 3-540-51515-1

Vol. 156: **Dyrberg, Thomas (Ed.):** The Role of Viruses and the Immune System in Diabetes Mellitus. 1990. 15 figs. XI, 142 pp. ISBN 3-540-51918-1

Vol. 157: **Swanstrom, Ronald; Vogt, Peter K. (Ed.):** Retroviruses. Strategies of Replication. 1990. 40 figs. XII, 260 pp. ISBN 3-540-51895-9

Vol. 158: **Muzyczka, Nicholas (Ed.):** Viral Expression Vectors. 1992. 20 figs. IX, 176 pp. ISBN 3-540-52431-2

Vol. 159: **Gray, David; Sprent, Jonathan (Ed.):** Immunological Memory. 1990. 38 figs. XII, 156 pp. ISBN 3-540-51921-1

Vol. 160: **Oldstone, Michael B. A.; Koprowski, Hilary (Eds.):** Retrovirus Infections of the Nervous System. 1990. 16 figs. XII, 176 pp. ISBN 3-540-51939-4

Vol. 161: **Racaniello, Vincent R. (Ed.):** Picornaviruses. 1990. 12 figs. X, 194 pp. ISBN 3-540-52429-0

Vol. 162: **Roy, Polly; Gorman, Barry M. (Eds.):** Bluetongue Viruses. 1990. 37 figs. X, 200 pp. ISBN 3-540-51922-X

Vol. 163: **Turner, Peter C.; Moyer, Richard W. (Eds.):** Poxviruses. 1990. 23 figs. X, 210 pp. ISBN 3-540-52430-4

Vol. 164: **Bækkeskov, Steinnun; Hansen, Bruno (Eds.):** Human Diabetes. 1990. 9 figs. X, 198 pp. ISBN 3-540-52652-8

Vol. 165: **Bothwell, Mark (Ed.):** Neuronal Growth Factors. 1991. 14 figs. IX, 173 pp. ISBN 3-540-52654-4

Vol. 166: **Potter, Michael; Melchers, Fritz (Eds.):** Mechanisms in B-Cell Neoplasia 1990. 143 figs. XIX, 380 pp. ISBN 3-540-52886-5

Vol. 167: **Kaufmann, Stefan H. E. (Ed.):** Heat Shock Proteins and Immune Response. 1991. 18 figs. IX, 214 pp. ISBN 3-540-52857-1

Vol. 168: **Mason, William S.; Seeger, Christoph (Eds.):** Hepadnaviruses. Molecular Biology and Pathogenesis. 1991. 21 figs. X, 206 pp. ISBN 3-540-53060-6

Vol. 169: **Kolakofsky, Daniel (Ed.):** Bunyaviridae. 1991. 34 figs. X, 256 pp. ISBN 3-540-53061-4

Vol. 170: **Compans, Richard W. (Ed.):** Protein Traffic in Eukaryotic Cells. Selected Reviews. 1991. 14 figs. X, 186 pp. ISBN 3-540-53631-0

Vol. 171: **Kung, Hsing-Jien; Vogt, Peter K. (Eds.):** Retroviral Insertion and Oncogene Activation. 1991. 18 figs. X, 179 pp. ISBN 3-540-53857-7

Vol. 172: **Chesebro, Bruce W. (Ed.):** Transmissible Spongiform Encephalopathies. 1991. 48 figs. X, 288 pp. ISBN 3-540-53883-6

Vol. 173: **Pfeffer, Klaus; Heeg, Klaus; Wagner, Hermann; Riethmüller, Gert (Eds.):** Function and Specificity of γ/δ T Cells. 1991. 41 figs. XII, 296 pp. ISBN 3-540-53781-3

Vol. 174: **Fleischer, Bernhard; Sjögren, Hans Olov (Eds.):** Superantigens. 1991. 13 figs. IX, 137 pp. ISBN 3-540-54205-1

Vol. 175: **Aktories, Klaus (Ed.):** ADP-Ribosylating Toxins. 1992. 23 figs. IX, 148 pp. ISBN 3-540-54598-0

Vol. 176: **Holland, John J. (Ed.):** Genetic Diversity of RNA Viruses. 1992. 34 figs. IX, 226 pp. ISBN 3-540-54652-9

Vol. 177: **Müller-Sieburg, Christa; Torok-Storb, Beverly; Visser, Jan; Storb, Rainer (Eds.):** Hematopoietic Stem Cells. 1992. 18 figs. XIII, 143 pp. ISBN 3-540-54531-X

Vol. 178: **Parker, Charles J. (Ed.):** Membrane Defenses Against Attack by Complement and Perforins. 1992. 26 figs. VIII, 188 pp. ISBN 3-540-54653-7

Vol. 179: **Rouse, Barry T. (Ed.):** Herpes Simplex Virus. 1992. 9 figs. X, 180 pp. ISBN 3-540-55066-6

Vol. 180: **Sansonetti, P. J. (Ed.):** Pathogenesis of Shigellosis. 1992. 15 figs. X, 143 pp. ISBN 3-540-55058-5

Vol. 181: **Russell, Stephen W.; Gordon, Siamon (Eds.):** Macrophage Biology and Activation. 1992. 42 figs. IX, 299 pp. ISBN 3-540-55293-6

Vol. 182: **Potter, Michael; Melchers, Fritz (Eds.):** Mechanisms in B-Cell Neoplasia. 1992. 188 figs. XX, 499 pp. ISBN 3-540-55658-3

Vol. 183: **Dimmock, Nigel J.:** Neutralization of Animal Viruses. 1993. 10 figs. VII, 149 pp. ISBN 3-540-56030-0

Vol. 184: **Dunon, Dominique; Mackay, Charles R.; Imhof, Beat A. (Eds.):** Adhesion in Leukocyte Homing and Differentiation. 1993. 37 figs. IX, 260 pp. ISBN 3-540-56756-9

Vol. 185: **Ramig, Robert F. (Ed.):** Rotaviruses. 1994. 37 figs. X, 380 pp. ISBN 3-540-56761-5

Vol. 186: **zur Hausen, Harald (Ed.):** Human Pathogenic Papillomaviruses. 1994. 37 figs. XIII, 274 pp. ISBN 3-540-57193-0

Vol. 187: **Rupprecht, Charles E.; Dietzschold, Bernhard; Koprowski, Hilary (Eds.):** Lyssaviruses. 1994. 50 figs. Approx. 280 pp. ISBN 3-540-57194-9

Vol. 188: **Letvin, Norman L.; Desrosiers, Ronald C. (Eds.):** Simian Immunodeficiency Virus. 1994. 37 figs. X, 240 pp. ISBN 3-540-57274-0